OUR UNIVERSE:

AN ARMCHAIR GUIDE

OUR UNIVERSE:

AN ARMCHAIR GUIDE

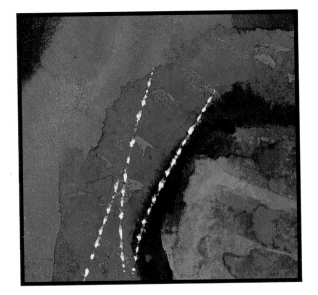

MICHAEL
ROWAN-ROBINSON

W. H. FREEMAN AND COMPANY
NEW YORK

Library of Congress Cataloging-in-Publication Data

Rowan-Robinson, Michael
   Our universe: an armchair guide / by Michael Rowan-Robinson.
     p. cm.
   Includes bibliographical references and index.
   ISBN: 0–7167–2156–2
   1. Astronomy. I. Title.
QB43.2.R675 1990
520 – dc20   90–36905
                                                  CIP

Published in the United States in 1990 by W. H. Freeman and Company, 41 Madison Avenue, New York, NY 10010. First published in the United Kingdom in 1990 by Longman Group UK Limited.

To Mary, Adam, Jonathan and Nicola

CONTENTS

PREFACE

When did we stop looking at the night sky? Today there are very few people, and that includes the astronomers, who can identify more than a few of the most obvious constellations. Yet it was once very different. Two or three thousand years ago, people had a thorough knowledge of the night sky and its motions. This is very clear from the many astronomical references in ancient literature, whether it be the Chinese lyric poets or the classical Greek and Roman writers. For Dante, astronomy was central to his whole vision of the world. The writings of Shakespeare and his contemporaries are full of astronomical allusions. But it is much more unusual to find astronomical references in modern writing.

This apparent indifference to the starry night does not just date from the moment when everyone uprooted and made for the smoke-shrouded cities. Surely the great Romantics, Goethe, Byron, *they* must have known the night sky? Not a bit of it. They talk airily of 'the stars' but it's hard to find much evidence that they knew the name of a single one. Since the sixteenth century, writers who have been familiar with the night sky – for example Milton, Tennyson, Thomas Hardy, Italo Calvino – have been the exception rather than the rule. But of course the stars, the universe, have remained a powerful image in all centuries.

I suppose people stopped looking at the sky because with Copernicus and Galileo everything was explained. The erratic motions of the planets along the zodiac held no more mystery, no longer was it rational to imagine that comets were terrible omens. Human destiny was not tied to the night sky and so we gradually stopped looking at it. But the discoveries of modern astronomy show that both our history and our destiny *are* bound up with the stars. The astronomers of the present age, with their superb telescopes on the tops of mountains or orbiting the earth on spacecraft, have given us images of the stars and galaxies unimaginable to the ancients. And with these images has come a new insight into the nature of the stars and of the universe.

Of course there are the armchair astronomers, those who are only too willing to study and discuss the night sky provided it is from the comfort of their armchairs. You are an armchair astronomer. But then so am I, along with most of that band of eccentrics who earn their living from astronomical research. We fly out to the exotic places where we like to put our telescopes and we pass the night in a warm room, sitting in an armchair, glancing at a television monitor, while a huge lump of metal and wires revolves in its dome nearby.

In this book I try to combine that wonder at the night sky that we have inherited from the past with the magnificent images and startling insights of modern astronomy. I outline the main ideas and theories of contemporary astronomy in a style intended to be accessible to someone with little or no scientific background. To relate this to our own direct experience of the night sky and to our cultural inheritance from the past, I have focused on twenty famous astronomical objects spanning the whole range of what there is in the universe, from comets to quasars. More than half of them are visible to the naked eye, have been known since antiquity, and are the source of mythological stories from many cultures and of literature from all ages. However their true nature has been revealed only in this century and, in many cases, only in the past twenty years. All the rest, except one (the quasar 3C273), have been known for at least two hundred years and can be seen with good binoculars or a small telescope.

Through these twenty objects I unfold what we know about the universe we live in, illustrating the narrative with a profusion of images from the world's ground-based and space-borne telescopes, a few famous paintings with an astronomical theme, and with quotations from the literature of past and present. In this way I have attempted to make modern astronomy and the once-familiar night sky more accessible to a wider audience.

When I heard the learn'd astronomer,
When the proofs, the figures, were ranged in columns before me,
When I was shown the charts and diagrams, to add, divide, and measure them,
When I sitting heard the astronomer where he lectures with much applause in the lecture
      room,
How soon unaccountable I became tired and sick,
Till rising and gliding out I wander'd off my myself,
In the mystical moist night-air, and from time to time,
Look'd up in perfect silence at the stars.

WHITMAN *By the Roadside*

Acknowledgements

I owe a considerable debt to Robert Burnham's excellent compilation 'Celestial Handbook', from which I conceived the idea for this book during a stormy week at the Roque de los Muchachos Observatory on La Palma. I would like to thank the many astronomers and observatories who have supplied me with the illustrations, particularly Jocelyn Bell-Burnell, Rosanne Hernandez, Peter Hingley, David Hughes, Christine Jones, Ian McLean, Patrick Moore, Paul Murdin, Fred Seward, Janet Sutherland, Peter Wilkinson and Iwan Williams. Anthony Rudolf generously took time off from his own book to read and comment on an early draft. My wife, Mary, read the manuscript and suggested many improvements. My thanks to her for that and everything.

Michael Rowan-Robinson, London   January 1990

COSMIC LANDSCAPE

Once it was the navigators crossing the oceans to find
new continents and new creatures, the globe opening up
before their eyes, and at the same time the unknown areas,
white on the map, shrinking.

Now it is the astronomers' telescopes penetrating the void
to find new worlds, voyages of discoveries made with giant
metal eyes, seeing light we cannot see.

No more the San Gabriel, the Santa Maria, the Victoria,
the Beagle or the Challenger. But Mount Palomar, Kitt Peak,
Green Bank, Mauna Kea, Medicina, Effelberg, Coonabarabrand.

Our voyages are made on a photon's back. Before us lies
the cosmic landscape, and our goal is nothing less than
the origin of life, earth, sun, stars, galaxies, universe.

C H A P T E R 1

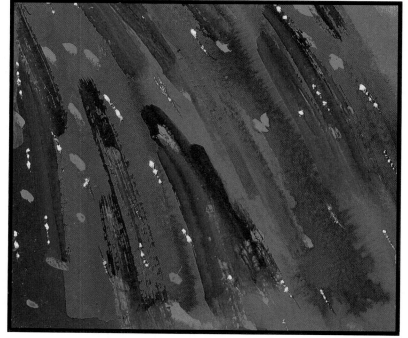

A comet arose whose body
was bright like the day,
while from its luminous
body a tail extended, like
the sting of a scorpion.

BABYLONIAN
INSCRIPTION 1140 BC

COMET HALLEY

Comets have been a source of mystery from the dawn of civilization to the present day. Halley's comet, which passed us by rather unimpressively in 1985–6, is the archetype of them all. Throughout recorded history it has displayed itself every seventy-six years to people across the globe. Most have a chance of seeing it once – a very few may see it twice. It was recorded many times in the annals of the ancient Chinese astronomers. They certainly recorded every apparition from 240 BC onwards, and may have seen it in 467 BC. It appeared in 1066 at the time of the Battle of Hastings and was stitched into the Bayeux Tapestry. The 1301 apparition figures in Giotto's fresco of the Adoration of the Magi in the sublime Scrovegni chapel in Padua. And in 1682 Edmund Halley produced the

▷ Fig 1.1 The return of Comet Halley in 1985-6. This photo was taken with the UK Schmidt Telescope in Australia on 12 March 1986

crowning triumph of Newton's theory of gravitation when he calculated its orbit, estimated its period as seventy-six years and identified several past apparitions:

> Now many things lead me to believe that the comet of the year 1531, observed by Apian, is the same as that which in the year 1607 was described by Kepler and Longomontanus, and which I saw and observed myself at its return in 1682 . . . The identity of these comets is confirmed by the fact that in 1456 a comet was seen, which passed in a retrograde direction between the Earth and Sun, in nearly the same manner, and although it was not observed astronomically, yet from its period and path I infer that it was the same comet as that of the years 1531, 1607 and 1682. I may, therefore, with some confidence predict its return in the year 1758. If this prediction is fulfilled, there is no reason to doubt that other comets will return.

Halley's comet is one of many hundreds which have plunged in from the outermost reaches of the solar system towards the sun, and have been woken from icy slumber into a blazing display of bright head and tail trailing across the sky. The awesomeness of a bright comet's gradual appearance terrified our ancestors and seemed to be a portent of disaster. Until the time of the Renaissance they were thought, under the powerful influence of Aristotle, to be an atmospheric phenomenon. Then in 1577 Tycho Brahe made careful observations of the comet of that year and showed that not only was it several times as far away as the moon, but also that it had actually passed through the zones where Aristotle's crystalline spheres were supposed to bear the planets round in their orbits. In 1531 Peter Apian had noted that comet tails always point away from the sun, an observation that had been made earlier by Seneca ('the tails of comets fly from the sun's rays') and by Chinese astronomers in AD 837. Studying Halley's comet in 1607 and the comet of 1618, Johannes Kepler concluded that the tail of a comet was pushed out of the comet's head by sunlight.

 After the triumphant success of Halley's prediction, comets began to be studied more systematically. Actually it is not such a simple matter to predict the return of a comet, because its orbit is modified by the attraction of the planets, especially Jupiter. The perturbations to cometary orbits caused by Jupiter and Saturn were calculated by Alexis Clairaut (1713–65), Pierre Simon Laplace (1749–1827) and William Olbers (1758–1840). In 1835 Friedrich Bessel explained irregularities in the time of return of Halley's comet as a rocket effect, after he had noticed a sunward plume of material like a blazing rocket. Biela's comet, which had been observed by the French philosopher Montaigne in 1726, was observed to split in two in 1846 and then return as two comets in 1852.

▷ *Fig 1.2 (inset) Comet Halley and Saguaro cacti seen from Northern Mexico, 23 Mar 1986*

▷ *Fig 1.3 Comet Halley passes near the densest portion of the Milky Way, almost in line with the centre of our Galaxy, in this photo taken at Cerro Tololo Observatory on 14 April 1986*

Old men and comets have been reverenced for the same reason; their long beards, and pretences to foretell events.

JONATHAN SWIFT
Thoughts on various subjects (1706)

◁ *Fig 1.4 Detailed structure in the tail of Comet Halley, 12 Mar 1986*

▽ *Fig 1.5 The passages of Comet Halley through human history*

−467	may have been recorded by Chinese astronomers
−391	
−316	
−240	definite Chinese record of every apparition from this date
−164	mentioned in Babylonian tablet (a)
−87	
−10	
67	
144	
221	
298	
375	
452	
529	
606	
684	recorded in Nuremburg Chronicles (b)
760	
837	
914	
990	
1066	appeared in the Bayeux tapestry (c)
1145	illustrated in the Eadwine Psalter (d)
1223	
1301	appears as the Star of Bethlehem in Giotto's fresco 'Adoration of the Magi', in the Scrovegni chapel in Padua (e)
1379	
1456	path sketched on contemporary star map (f)
1531	studied by Peter Apian, who recognized that comet tails point away from the sun (g)
1607	studied by Kepler and Longomontanus
1682	observed by Edmund Halley (h), who calculated its orbit. The contemporary print records its appearance from Nuremburg (i)
1759	Halley's prediction of return of comet confirmed. The comet features in this painting by Samuel Scott (j)
1835	The head of the comet sketched by the German astronomer Friedrich Bessel (k)
1910	first photographic record (l)

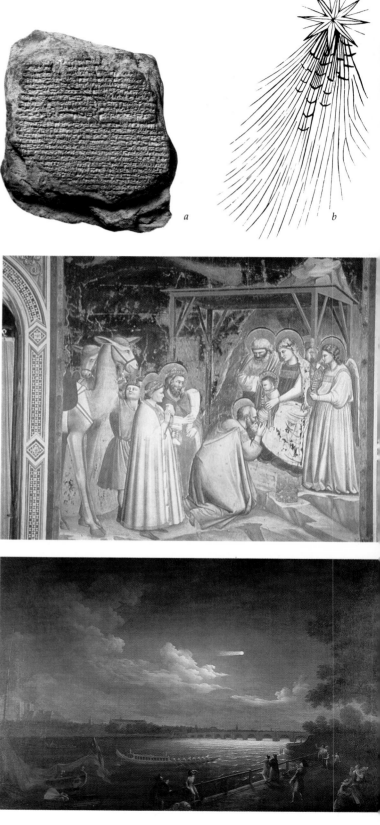

a *b*

c

d

HEBR.   ROM.   GALL.

f

g

Dr Halley

h

i

k

l

During the present century, astronomers have gradually unravelled the mystery of comets. Comets are the relics of the primordial material of the solar system, from which the sun and planets formed, and consist, according to a picture first proposed by Fred Whipple in 1950, of aggregates of rock, dust and ice, perhaps twenty miles across. Thousands of millions of comets orbit the sun in a cloud far beyond the orbit of the last planet. Usually the last planet is Pluto, but because this has a strongly elliptical orbit, every 248 years it dips inside Neptune's orbit, as it did in 1979. Until 1999 Neptune will be the outermost planet. This cloud of comets, named after the Dutch astronomer Jan Oort who first deduced its existence also in 1950, lies about 1000 times further away than Pluto, or 50,000 times further away than the sun. The sun's gravitational grip on the Oort Cloud is therefore tenuous and the comets also respond to the gravitational attraction of passing stars and of nearby vast clouds of interstellar gas.

When some accidental gust of gravity deflects one in towards the sun, it embarks on an elongated, elliptical orbit reaching from the far depths of the solar system down to the vicinity of the sun. As the comet arrives near the sun, the ices on the surface melt and are vaporized, and gas and dust particles stream off the core, or *nucleus*, of the comet in two separate tails pointing away from the sun. The straighter tail is due to hot molecules of gas, the curved tail is due to particles of dust. Comets' tails can extend to millions of miles in length. This gradual whittling away of a comet's dust and ice each time it approaches the sun means that after perhaps a few hundred apparitions nothing will be left but lumps of rock, which hurtle by us almost unnoticed.

Some tens of comets are sighted each year, many by amateur astronomers. Quite a high proportion of these sightings turn out to be the return of a comet already known. The last really bright comet, which could be seen in the daytime, was the Great Comet of 1910. This was much brighter than Halley's Comet, which also passed by that year. During that passage the earth passed right through the tail of Halley's comet without apparent mishap. Comets bright enough to be seen with the naked eye at night (if you know where to look) occur about every ten years.

△ *Fig 1.6 Other comets (above and right)*
a *painting of Comet Donati 1858*
b *'Pegwell Bay', by William Dyce, with comet Donati in the background (top centre)*
c *Comet Bennett 1970*
d *Comet Mrkos 1957*
e *Comet West 1976*

△ *Fig 1.7 Aztec painting depicting the fall of Moctezuma*

△ *Fig 1.8 An 1857 cartoon satirizing contemporary fear of come*

b

c

d

e

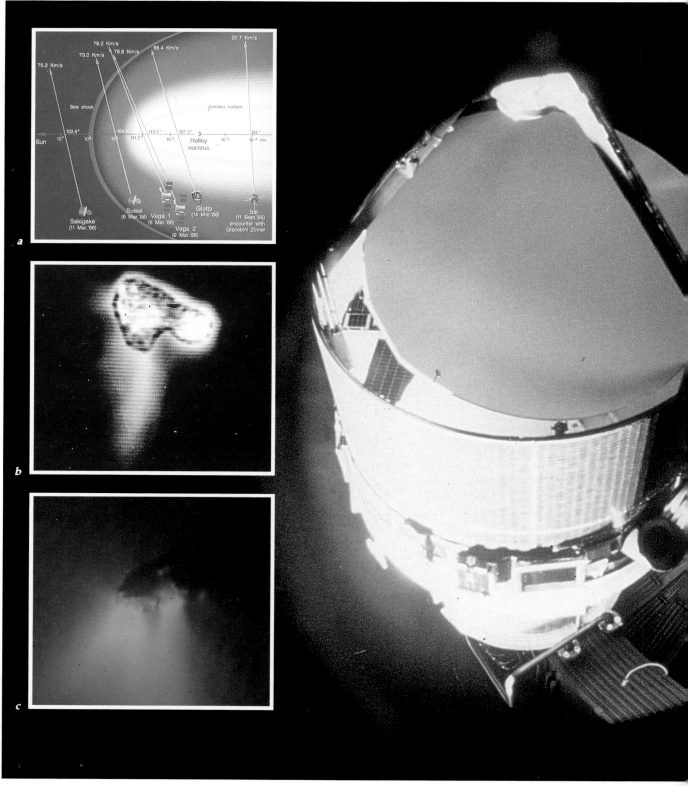

a

b

c

d

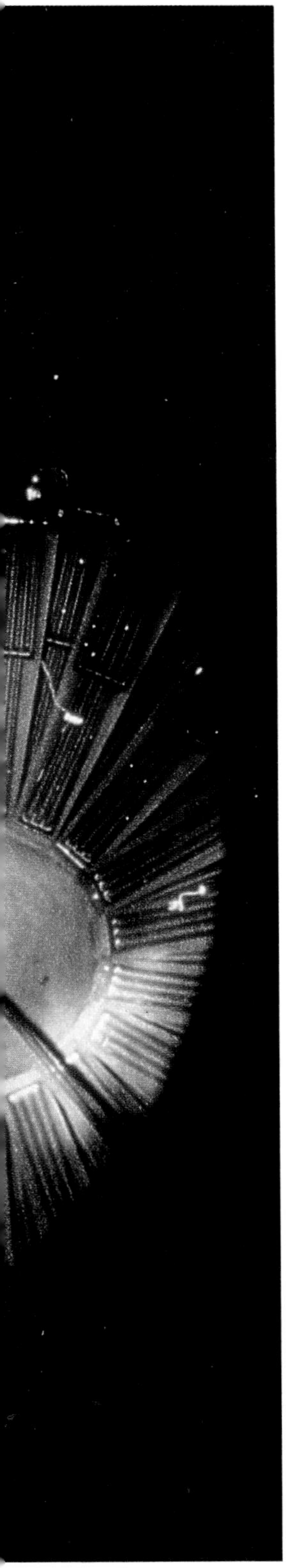

When Comet Halley returned to our neck of the woods in the winter of 1985–6, astronomers were ready with an armada of spacecraft, culminating in the passage of the European mission Giotto close to the nucleus on 13 March 1986. Giotto's television camera recorded this striking image of the heart of Comet Halley from a distance of two thousand miles. However as the camera had been programmed to point at the brightest part of the comet, at the actual moment of closest passage they were pointing *away* from the nucleus towards a brilliant gas-jet spurting out of the core. Hence the rather confused live broadcast of the encounter, which many of us watched in some perplexity.

It was another comet of the 1980s, known as Comet IRAS 1983n (comets are labelled by the name of their discoverer, in this case the IRAS Infrared Astronomical Satellite, the year of their discovery, and a letter of the alphabet), which established the link between comets and three other fascinating phenomena: *Apollo asteroids, meteorites* and *meteors*. Asteroids are lumps of rock ranging in size from one to hundreds of miles across and most of them are circling the sun in a ring between Mars and Jupiter, the *asteroid belt*. However some of them have orbits more like comets and cross the earth's orbit. These are the *Apollo asteroids*. Luckily most of them cross the earth's orbit when the earth is somewhere else, but occasionally one does hit the earth and the consequences can be dramatic. Such collisions were much more common in the early days of the solar system and an impression of their effect can be seen in the form of the vast craters frozen into the moon's surface. The corresponding giant craters left on earth have been eroded away by wind and rain. It may have been an Apollo asteroid which finished off the dinosaurs and many other species by throwing up a huge cloud of dust and turning day to night for months on end. The origin of these menacing cosmic wanderers is a matter of more than passing interest to us.

Meteorites are lumps of rock, iron or coal-like (carbonaceous) aggregates which crash to earth from the sky. They have been venerated throughout history. The Kaaba, the sacred black stone at Mecca, the centre of Islam, is almost certainly a meteorite, and a similar stone was venerated by the Pawnee Indians of Nebraska. Yet it took astronomers many centuries to believe in the reality of stones from the sky. Even though thinkers from the Renaissance onwards recognized that celestial bodies were likely to be made of the same material as the earth, they still seem to have been trapped in some Platonic idea of the inviolability of the earth. In 1803 a commission of experts was sent by the French Academy of Sciences to the town of L'Aigle to investigate reports that thousands of stones had fallen on the town from the sky. They came away reluctantly convinced. It took that rarest of events, the disintegration of a meteorite over a populated area, to persuade scientists of the truth of what had been known to many isolated individuals throughout history.

◁ *Fig 1.9 Space-craft encounters with Comet Halley 1986*
a The armada of space-craft which sailed to Halley
b View of Comet Halley nucleus from the Soviet VEGA-2
c View of Comet Halley nucleus from the Multicolour Camera on board ESA's Giotto spacecraft, at a range of only a few thousand kilometres
d ESA's Giotto spacecraft undergoing pre-launch tests

▷ *Fig 1.10 Two meteorites, perhaps once part of a comet nucleus*

Meteors, on the other hand, are particles of dust flashing through the earth's atmosphere, which heat up until they become incandescent and glow as a 'shooting star'. Meteors, or shooting stars, are seen especially frequently on particular nights of the year. I remember sleeping in the open on a summer night in France and seeing many hundreds, the famous Perseids. In 1866 Giovanni Schiaparelli realized that meteor streams occur when the earth passes through the orbit of a dying comet which has left its debris strewn along the way. Most of the impressive meteor streams are associated with known comets. For example, the August Perseids move in the same orbit as Comet Tuttle, the November Leonids follow that of Comet Tempel and the May Aquarids are associated with Comet Halley itself.

The Infrared Astronomical Satellite, IRAS, was launched in January 1983 and surveyed the sky at infrared wavelengths for ten months. I was a member of the team of scientists who planned the mission and analysed the signals recorded by IRAS's infrared telescope and detectors, and transmitted back to earth. IRAS found several comets. The first, IRAS-Iraki-Alcock, was especially exciting because it passed the closest to earth for several hundred years. At three million miles, it passed at only ten times the distance of the moon. When the orbit of the comet was first calculated it was not certain it would miss the earth. Our best estimate was that it would miss, but we could not be sure. We sat round debating whether we should warn the world's population of this worrying fact. Personally I thought we should, because if it did hit some large city like London, Tokyo or New York and we had failed to warn of this we were likely to be unpopular, to say the least. Typical scientific caution prevailed, however, and we waited a few more days to get a better estimate of the orbit. The danger of a collision receded. The comet became visible to the naked eye as a faint blur of light before passing on its way.

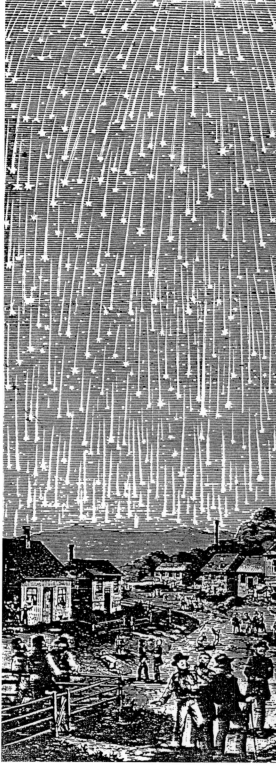

▷ *Fig 1.11 Photograph of Comet IRAS-Iraki-Alcock, which passed only 3 million miles from the earth, taken from Kitt Peak Observatory on 8 May 1983. The trailing of the stars indicates how far the comet moved during this 10-minute exposure*

▷ *Fig 1.12 False-colour picture of infrared emission from Comet IRAS-Iraki-Alcock. The nucleus of the comet was 6 miles in diameter and if it had hit the earth it would have made a crater one hundred miles across*

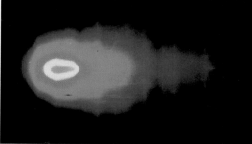

△ *Fig 1.13 Historical woodcut of the Leonids meteor shower*

THE WAVELENGTH BANDS

It is only in the past century that most people have begun to be aware of invisible radiations. Today everyone has heard of radio waves and X-rays, though probably few people realize that they are the same kind of thing as the visible light with which we see our world. The first of the invisible radiations to be discovered was infrared radiation, which was discovered accidentally by William Herschel in 1800. He was studying sunlight after it had passed through a prism and been broken up into the colours of the spectrum: red, orange, yellow, green, blue, indigo, violet. Testing the different colours with a thermometer, he noticed that the thermometer soared up beyond the red end of the spectrum. He had shown that radiant heat is a kind of light like red light but happens to be invisible to the human eye, and he called it *infrared* radiation. Soon after, invisible light beyond the violet end of the spectrum, *ultraviolet* light, was found through its effect on photographic emulsion.

The different colours which make up the visible spectrum differ from each other only in their wavelength, with red light having about twice the wavelength of violet light. Beyond the ultraviolet, at even shorter wavelengths, we come first to *X-rays* and then to the even more penetrating *gamma-rays*. At the longer wavelengths, beyond the infrared, we come to *microwaves* and then to *radio* wavelengths. The names of the different wavelength regions, or wave bands, are rather arbitrary, and arose mainly for historical reasons.

The energy carried by a particle of light, or *photon*, is proportional to the frequency of the light, so the sequence of wavebands – radio, microwave, infrared, visible, ultraviolet, X-ray, gamma-ray – is one of increasing energy per photon. This in turn means that we are generally looking at progressively hotter phenomena as we go from infrared to X-rays, say.

Astronomers use all of these wave bands, though for some we have to get our telescopes above the earth's atmosphere to see out, by using aircraft, balloons or, ideally, satellites. In the radio band we have to persuade broadcasters to leave a few wavelengths free of transmissions to avoid interfering with the very faint astronomical sources of radiation. Television and satellite communication links are beginning to cause us headaches in the microwave band. Perhaps we are the last generation of astronomers who can use all the wave bands.

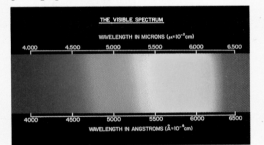

THE VISIBLE SPECTRUM

WAVELENGTH IN MICRONS (μ=10^{-4}cm)

| 4.000 | 4.500 | 5.000 | 5.500 | 6.000 | 6.500 |

| 4000 | 4500 | 5000 | 5500 | 6000 | 6500 |

WAVELENGTH IN ANGSTROMS (Å=10^{-8}cm)

The wavebands of modern astronomy	wavelength												
	100m	1m	1cm	100μm	1μm	10nm	10^{-1}nm	10^{-3}nm		10^{-5}nm	10^{-7}nm		
	radio		microwave	submillimetre	far infrared	near infrared	visible	ultraviolet	soft X-ray	hard X-ray		γ-ray	
	6	8	10	12	14	16	18	20		22	24		
	1g (frequency in Hz)												

△ *(top) The colours of the visible spectrum correspond to different wavelengths, from violet (short wavelengths) to red (long wavelengths), but the wavelengths visible to the human eye represent only a tiny portion of those now used by astronomers.*

△ *The wavelengths of the electromagnetic spectrum are arbitrarily divided into wavebands, with names which relate to how they were discovered or how they are detected. The main wavebands are: radio, microwave, submillimetre, far infrared, near infrared, visible, ultraviolet, soft X-ray, hard X-ray and gamma-ray.*

The last of IRAS's comets, 1983n, turned out to have the same orbit as the Geminid meteor stream. Moreover its appearance was that of an Apollo asteroid, a dead lump of rock with no gas or dust tails: it was given the name Phaethon. Thus at one blow it was demonstrated that at least some of the Apollo asteroids are simply the nuclei of dead comets which have exhausted their envelope of dust, gas and ice. Many meteorites too are almost certainly cometary debris. Detailed analysis of their composition shows that others are fragments from collisions between asteroids in the asteroid belt, and still others are fragments chipped off the moon or other planets by asteroids. For several comets, notably Comet Tempel, IRAS was able to map the trail of debris spread out along the orbit of the comet.

Yet another insight into the violent life of the solar system came with the IRAS team's discovery of bands of dust spread round the sky close to the plane of the ecliptic. The *ecliptic* is the plane of the earth's orbit around the sun and, approximately, of the orbits of the planets and asteroids. These bands turned out to be the debris of collisions between members of families of asteroids which move in similar orbits within the asteroid belt.

Meteorites give us a glimpse of the primordial material from which the planets of the solar system formed. A small portion, or *inclusion*, of the Allende meteorite, which crashed to earth near the village of Allende in Mexico in 1969, can be shown to have formed containing an abnormally high proportion of a radioactive form of aluminium (Aluminium-26) with a half-life of only a million years. The half-life is the time in which, on average, half the atoms of a radioactive element have been transformed by emitting a beta-ray (an electron moving at very high speed). Most naturally occuring radioactive elements on earth today take thousands of millions of years to decay, like Uranium-235 (from these we can deduce the age of the earth and moon to be 4.5 thousand million years). Now most radioactive elements are made, as we shall see later, in supernova explosions, the death-throes of massive stars. This little fragment of the Allende meteorite tells us that a supernova exploded nearby very shortly before the solar system formed. Was there a connection? Could the supernova explosion have triggered the collapse of the cloud of dust and gas from which the solar system formed?

Meteorites come in three kinds: stony, iron or carbonaceous. The last of these, with a crumbly, coal-like consistency, are the most bewildering because a wealth of complex, carbon-based ('organic') molecules have been found in them, even including many of the amino-acids which form the basis of the chemistry of living things. This has led Fred Hoyle and Chandra Wickramasinghe to speculate that life came to the earth via comets. Certainly the dense cloud of gas from which the solar system formed was the scene of intense chemical activity. But probably any large organic molecules around would have been destroyed during the forma- tion of the earth. Meteorites landing on the earth subsequently could indeed bring organic molecules with them. But far more important would seem to be the evolution of an environment in which self-replicating molecules like DNA could emerge. Few astronomers or biologists take Hoyle and Wickramasinghe's 'Life Cloud' very seriously. The message of the carbonaceous chondrite meteorites is that the first steps along the road to life, formation of complex molecules like amino-acids, is easy. The mystery is how the next step took place.

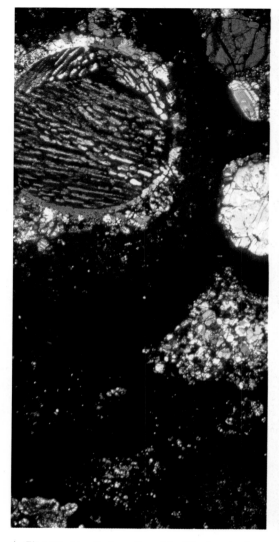

△ *Fig 1.14 A 'section', or slice, of the Allende meteorite*

Satan stood
Unterrified, and like a comet burned
That fires the length of Ophiuchus huge
In th'Arctic sky, and from his horrid hair
Shakes pestilence and war.

MILTON *Paradise Lost*

When beggars die, there are no comets seen;
The heavens themselves blaze forth the death of princes.

SHAKESPEARE *Julius Caesar*

△ Fig 1.15 *The Barringer meteorite crater in Arizona, which is a mile across*

▷ Fig 1.16 *The dust bands running round the plane of the ecliptic discovered by IRAS. These are believed to be the debris of collisions between members of asteroid families with similar orbits, located in the asteroid belt between Mars and Jupiter*

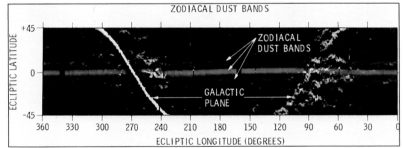

C H A P T E R 2

When the night drives away the limpid day
And deep gloom brings to others the dawn
I gaze in wonder at the cruel stars.

PETRARCH *Sonnets*

ALPHA CENTAURI

NEAREST STAR TO THE SUN

The brightest star in the southern hemisphere constellation of the Centaur, called Rigel Kentaurus by the medieval Arab astronomers, is in fact the closest star system to the sun. The sun is of course the closest star to earth itself. To us this seems commonplace, but it was a very great discovery of the Renaissance that the sun is just a star like any other. Most of the ancient thinkers, whether Greek, Babylonian, Egyptian or Chinese, believed that there was a fundamental distinction between the material of the earth and that of the heavens, and between the sun and the stars.

When we look at the terrestial landscape and see the sun blazing down, it is hard to believe that the sun and earth are composed of the same material, the same elements. The sun is a very hot ball of gas and its physical conditions are more drastic than those created anywhere on earth until the advent of that monstrous human creation, the H-bomb. The temperature of the surface of the sun is over 6000 degrees centigrade and of the centre, several million degrees. Even so, all the elements detected on the sun are present on earth too. In the nineteenth century the element helium was first discovered during observations of the sun (see p. 54), but it too was eventually found on earth.

At night when we look up at the stars it is even harder to imagine their connection to the earth. The sky appears so dark and cold, the stars so feeble and distant. The pale moon alone seems to occupy an intermediate realm. Indeed it was Galileo's first glimpse of the moon through a telescope which shattered the concept of a distinct celestial substance. For he saw that the moon had mountains on it, that it too was a landscape.

Alpha Centauri occupies a special place in the heavens for us humans. It is only 4.3 light years away, or 25 million million miles. Such a distance, of course, takes the breath away. The moon is a quarter of a million miles away, the sun 93 million miles. Light takes eight minutes to travel from the sun to earth. From Alpha Centauri the light takes over four years to get

here, travelling at 186,000 miles per second. And yet, as we shall see, the distance to Alpha Centauri is as nothing compared with the furthest reaches of the universe that we can survey today. To voyage through the astronomical landscape you have to abandon all thought of miles. I remember that, when I began to study astronomy in my early twenties, the first months were spent in a total reorganization of my mental attitudes to distance, so that I could start to think of the step to Alpha Centauri as a minuscule one. For me, these two worlds, that of the one-mile walk to the station or the three thousand mile plane journey across the Atlantic, and that of the ten thousand million light years or sixty thousand million million million miles to the most distant galaxy yet known, these two worlds coexist uneasily in my head. Yet all the time I somehow want to bridge them, I want to feel the universe as a landscape in which we exist. If I am honest, I can do this sitting in my armchair, imagining, but I cannot do it while I am actually looking up at the stars.

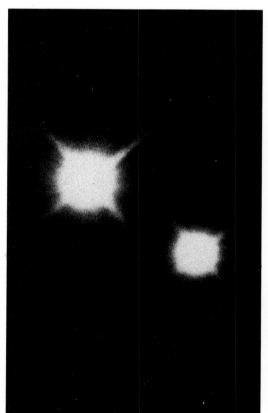

The ancient Greek astronomers, who knew so much, realized that the stars must be very distant. Nicolas Copernicus, when he discovered that the earth orbits round the sun, realized that the stars must be exceedingly distant compared even to the sun, otherwise the nearer ones would appear to change direction as the earth traced out its orbit. Yet it was not till 1839, almost three hundred years after the death of Copernicus, that the distance to Alpha Centauri was finally measured, by Thomas Henderson, from analysis of observations made at the Cape of Good Hope Observatory in South Africa. (This observatory, established by British settlers, remained an enclave of British astronomy right up till 1982, when the UK astronomy community was persuaded that it was shameful to be maintaining the last official links with apartheid.)

◁ *Fig 2.1 Part of the Centaurus and Carina constellations, showing Alpha and Beta Centauri (upper left), the Southern Cross (just above the centre), and Eta Carina (the fuzzy object to the lower right)*

THE CONSTELLATIONS

The division of the bright stars into constellations has been made several times in human history. The ancient Chinese astronomers divided the stars into some seven hundred 'asterisms', all a good deal smaller than the constellations we use today. At the other extreme, the Indians of Central America see the figure of a giant covering almost the whole of the sky. The constellations we use today, their names and the stories about them, have come down to us from the Greeks. From the area of the sky covered by these ancient named constellations, it has been deduced that the choice of constellations was probably made by the Minoans of Crete.

The stars of a constellation like Orion or the Great Bear do not necessarily lie close together in space – they just happen to lie in similar directions in the sky.

The constellations visible from the northern hemisphere according to Peter Apian in 1540.

◁ *Fig 2.2 Close-up of the double star Alpha Centauri, the nearest star to the sun and the third brightest star in the sky. This pair, along with the faint companion Proxima Centauri, or Alpha Centauri C, form the nearest star system to our own at a distance of only 4.3 light years. The photo shows Alpha Centauri A (the larger star) and B*

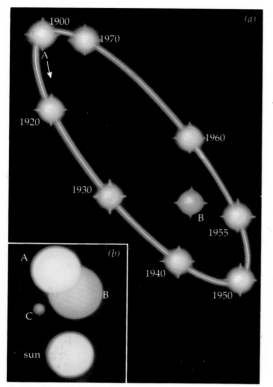

△ *Fig 2.3a The orbit of Alpha Centauri A around component B over an 80 year time-span*
b The relative sizes of Alpha Centauri A, B and C. A and B are similar to the sun in size and mass. Component C is very much smaller and is only about one tenth the mass of the sun

Henderson observed Alpha Centauri over the course of a year during 1832–3, and then returned to his native Edinburgh to analyse his observations. He saw immediately that Alpha Centauri changed its direction to and fro as the earth moved first to one side of its orbit around the sun and then to the other. Unfortunately it took him several years to complete his analysis and he did not announce his results until 8 January 1839. As a result of this delay, Henderson was not the first to announce this kind of apparent motion of a star on the sky, *parallax* as it is called by astronomers. The German astronomer Friedrich Bessel had published two months earlier the results of his studies of the obscure star 61 Cygni, which also showed parallactic motion, corresponding to a distance of around 10 light years (modern value 11.5 light years). Finally, after three hundred years, the Copernican system had been proved beyond a doubt.

The discovery of parallax was not the first breach in the concept of the 'fixed' stars. Edmund Halley, whom we encountered in the previous chapter, noticed in 1718 when he compared the positions of bright stars with those recorded by Hipparcos of Rhodes two thousand years earlier, that three stars (Sirius, Aldebaran and Arcturus) had noticeably changed their positions. The largest such motion is that of Alpha Centauri, in fact. Such motions of stars are called *proper motions* by astronomers. We now know that all the stars are swirling through the sky and it is only because the time-scale for the swirling is so long, a hundred million years, that we do not notice it. But for the nearest stars we can, with great care, with the devotions of lifetimes like those of Hipparcos and Halley, notice their motions.

> . . . the fixed stars
> **Are moving really and the whole Galaxy turning**
> **Round and round on its own axis agitatedly . . .**
>
> PETER RUSSELL *'Elegiac'*

PARALLAX

As we sit in a moving train, the distant parts of the landscape hardly change but the nearer parts change their appearance continuously as we view them from different angles. Similarly, as the earth orbits round the sun the stellar landscape changes its aspect, with the nearest stars changing their directions to and fro as the earth orbits back and forth. This is the phenomenon of *parallax*. The amount of the change in direction of a star over a six-month period, which takes the earth from one side of its orbit to the other, gives a measure of the distance of the star – the smaller the change, the more distant the star. Even for the nearest star (Alpha Centauri) the change is only 1.5 arcsec or 1/2400th of a degree, so its detection requires very accurate measurements with a large telescope.

With measurements from the ground, accurate parallaxes can be measured for stars up to about sixty light years away. The Hipparcos satellite, launched by the European Space Agency in August 1989, was supposed to reach ten times further away than that. Unfortunately it failed to achieve the correct orbit, so may not succeed in this. The Hubble Space Telescope, launched on 24 April 1990, will also measure the parallax of distant stars.

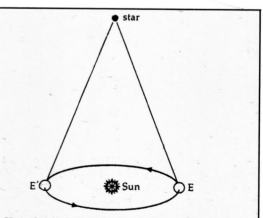

How the direction of a star changes as the earth orbit rounds the sun. This sketch is highly exaggerated. In fact the effect is so small that it was not measured even for the nearest stars until 1838, almost 300 years after the death of Copernicus, who first predicted the effect.

△ *Fig 2.4 A typical modern astronomical observatory with a series of domes each containing a telescope, on a remote mountain top. This is the US National Optical Astronomy Observatory on Kitt Peak mountain in the Arizona desert*

▽ *Fig 2.5 Motion on the sky of Alpha Centauri C over a 43-year period. It moved 2.75 arcminutes in this interval. It is two light months from Alpha Centauri A and B and will take half a million years to complete its orbit around them*

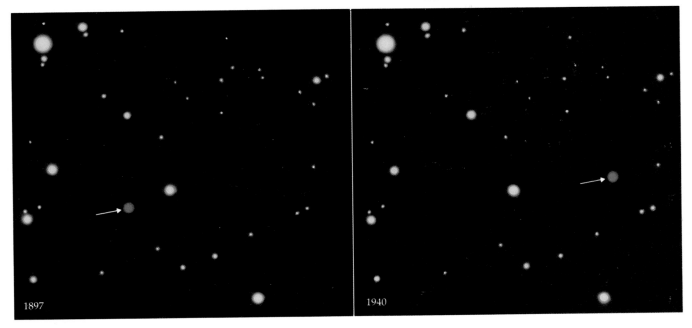

1897

1940

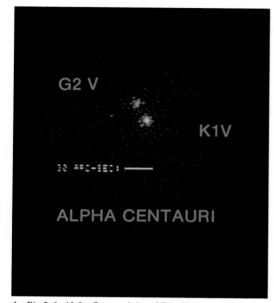

△ *Fig 2.6 Alpha Centauri A and B in X-rays*

Between the stars, how distant

RILKE *Sonnets to Orpheus*

They cannot scare me with their empty spaces
Between stars – on stars where no human race is.

ROBERT FROST *'Desert Places'*

Alpha Centauri differs from the sun in one major respect. It is not a single star, but in fact three in orbit round each other. The two brightest stars of the system, A and B, comprise one of the most impressive 'visual binary' systems in the night sky. Their double nature was discovered by Father Richaud at Pondicherry, India, in 1689, during his observations of the comet of that year. The two stars complete their orbit around each other every eighty years.

The third star of the system, Alpha Centauri C, or Proxima Centauri, was discovered only in 1915 by R. T. Innes, an Australian working at the Union Observatory, Johannesburg. It is one of the least luminous and least massive stars known. It radiates only 1/13000 times as much light as the sun and is only about one tenth as massive as the sun. In fact from theoretical arguments we know that if it were any less massive than this it would not be a star at all, but only a 'brown dwarf', doomed never to switch on the thermonuclear power station at its core which powers a star.

Alpha Centauri C is unusual in another way. It is one of the most violent 'flare' stars known. Flares have been noticed on the sun for thousands of years. Every so often a spot on the sun brightens suddenly. This is accompanied by a violent electromagnetic outburst, causing radio and X-

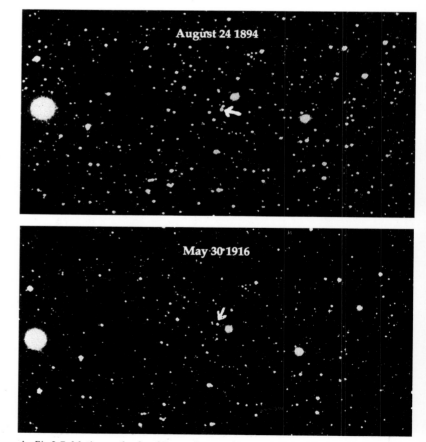

△ *Fig 2.7 Motion on the sky of Barnard's star, the star which is changing its position on the sky most rapidly of all*

ray storms and accelerating atomic particles to speeds close to that of light. Astronauts could get fried by these high-energy particles if they were caught aloft during a solar flare. The large dose of high-energy particles from solar flares and other more distant sources which would be experienced by interplanetary travellers may well prove an unacceptable price for the piloted missions to Mars proposed by the Soviet Union and United States for the next century. Even to be in Concorde at its cruising altitude during a solar flare would not be good for your long-term health. The frequency of solar flares vary in an eleven-year cycle, being rare at solar minimum but more common (several per year) at maximum, eleven years later. Puny Alpha Centauri C, however, had at least forty-eight flares between 1925 and 1950, as Harlow Shapley discovered from careful observations, each flare being as powerful as those rarer events on its enormously more luminous neighbour, the sun.

Could Alpha Centauri have planets like our solar system? We do not know yet. It is possible that only single stars like the sun have planets. Multiple stars, which are the rule rather than the exception, may not be the place to look for other earths. Planets in Alpha Centauri could have very complicated orbits, passing first one star and then another in the system. It seems unlikely that life could survive such variable conditions.

△ *Fig 2.9 A flare explodes on the surface of the sun*

The radiance of the star that leans on me
Was shining years ago. The light that now
Glitters up there my eyes may never see,
And so the time lag teases me . . .

ELIZABETH JENNINGS *'Delay'*

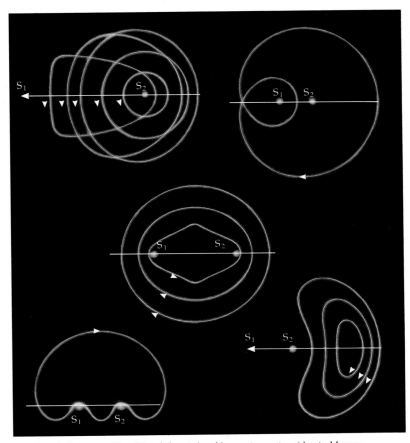

△ *Fig 2.8 Some possible orbits of planets in a binary star system (due to Murray Schecter). The two stars are labelled S_1 and S_2*

C H A P T E R 3

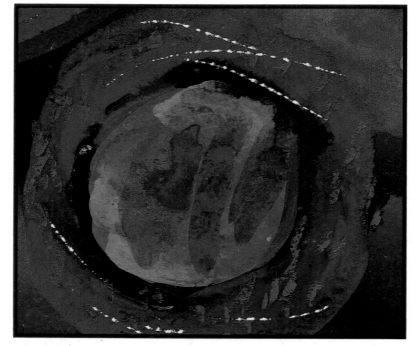

Gloomy Orion and the Dog
Are veiled . . .

T S. ELIOT *'Sweeney Among the Nightingales'*

SIRIUS AND ITS WHITE DWARF COMPANION

Sirius, 'the Sparkling One', 'the Scorching One', or 'the Dog Star', is the brightest star in the sky, 'the leader of the host of heaven'. It dominates the winter night sky in the northern hemisphere. To Caroline and William Herschel at the turn of the nineteenth century, the approach of the star to the field of view of their great reflecting telescope was heralded by a glow resembling a coming dawn, and its actual entrance was almost intolerable to the eye. The colour of Sirius is white with a tinge of blue when it is high in the sky, though at lower altitudes it seems to flicker with all the colours of the rainbow in the scintillations of the earth's atmosphere.

◁ *Fig 3.1 Sirius rises. Sirius is to the left, the constellation of Orion is in the centre and Aldebaran (Alpha Tauri) is to the right*

In ancient Egypt, Sirius was revered as 'the Nile Star' or 'Star of Isis'. Its annual appearance just before dawn at the summer solstice announced the flooding of the Nile, upon which Egyptian agriculture – and indeed all life in ancient Egypt – depended. From as early as 3000 BC this 'heliacal rising' is referred to in temple inscriptions, where the star is identified with the soul of the goddess Isis. In the temple of Isis-Hathor at Dendera, which is oriented to the rising of Sirius, appears the inscription:

> Her Majesty Isis shines into the temple on New Year's Day, and she mingles her light with that of her father Ra on the horizon.

Ra was the Egyptian sun-god, so this inscription tells us that the Egyptian New Year began at the point when Sirius rises with the sun, its heliacal rising.

Sirius is the brightest star of the constellation Canis Major, the Great Dog. The association of Sirius with a celestial Dog goes back at least to the Babylonians. There is a striking passage in Homer's *Iliad* where King Priam, from the walls of Troy, watches the wrathful Achilles advancing across the Trojan plain

> ... blazing as the star that cometh forth at Harvest-time,
> shining forth amid the host of stars in the darkness of the night,
> the star whose name men call Orion's Dog.
> Brightest of all is he, yet for an evil sign is he set,
> and bringeth much fever upon hapless men ...

In the ancient Greek and Roman world the influence of Sirius was regarded as malign. Virgil in the *Aeneid* writes of 'the Dog Star, that burning constellation, when he brings drought and diseases on sickly mortals, rises and saddens the sky with inauspicious light'. The scorching heat of July and August, which occurs when Sirius rises with the sun, was attributed to the dire influence of the blazing star, bringing forth fever in people and madness in dogs. These ideas prevailed well up to the time of the Renaissance, for we find Dante speaking of 'the great scourge of the dog days'. Even as late as the eighteenth century Alexander Pope wrote, in jocular vein:

> Shut, shut the door, good John! fatigued I said,
> Tie up the knocker, say I'm sick, I'm dead
> The dog-star rages.

Sirius is a blue-white star about 23 times as luminous as the sun, about 1.8 times the diameter and about 2.35 times the mass. Its surface temperature is about 10,000 degrees centigrade. It is the fifth nearest star to the sun, at eight light years, the second nearest of the naked eye stars (after Alpha Centauri), and the nearest naked eye star visible from northern latitudes. Thus if you live at northern latitudes, where Alpha Centauri is never visible, and want to stare out at the nearest of those terribly distant stars, you should go out on a January or February evening, when Orion is at its

The kingly brilliance of Sirius pierced the eye with a steely glitter....

THOMAS HARDY *Far From the Madding Crowd*

That is how you should picture sun and moon and stars – as showering their splendour in successive outbursts and for ever losing flash after flash of flame, not as enduring essences untouched by time.

LUCRETIUS

△ *Fig 3.2 Representations of Orion and Sirius on an ancient Egyptian tomb-painting at Dendera. Orion travels through the heavens in his celestial boat, followed by Sirius, shown as a kneeling cow*

▷ *Fig 3.3 Three close-ups of Sirius. The spikes are artefacts caused by the telescope. In the upper photos the faint dot is Sirius's white dwarf companion, Sirius B*

highest to the south, and find Sirius blazing away to the lower left. Like Alpha Centauri, Sirius is not a single star, but has an extremely interesting faint companion, as we shall see below.

The energy of stars is generated by nuclear reactions in the core of the star, mainly through the fusion of hydrogen into helium. In the process, about 0.7 per cent of the mass of the core is converted into energy. We know the mass of the hot core of a star from detailed computer calculations, so we can work out how long the star will live. Now although Sirius is only just over twice the mass of the sun, it is squandering its energy resources at over twenty times the sun's rate. Thus Sirius will only live for one tenth as long as the sun. The sun is 4.5 thousand million years old and is about half-way through its life. To some this is a shocking idea. It's bad enough that our own lives are finite, but the idea that the sun will eventually die, and with it life in the solar system, is hard to face. Sirius, on the other hand, will live for less than a thousand million years in all. This is one of the most striking features of the stellar landscape, that the more massive stars are relatively transient.

There is a curious paradox about the colour of Sirius in ancient times which seems to suggest even more rapid change. Several classical writers, notably Cicero, Horace and Seneca, refer to Sirius as ruddy or red, and the great Alexandrine astronomer Ptolemy classified Sirius as 'fiery red'. Yet the Arab astronomer, Al Sufi, in his star catalogue prepared some eight hundred years later, around AD 925, fails to include Sirius among stars he calls red and it appears to have been white ever since. In the thirteenth-century collection of pastoral songs *Carmina Burana*, the whiteness of Sirius is compared to that of ivory-white teeth, and Geoffrey Chaucer in 1391 refers to Sirius (known to the Arabs as Al-Abur) as 'the faire white sterre that is clepid Alhabur'.

Could Sirius have changed colour in a mere eight hundred years? Or did the ancient writers simply make a mistake? The latter is not impossible since there are other stars classified by Ptolemy as 'fiery red' which we would today call yellow. A star like Sirius should only change its appearance over hundreds of millions of years. Even when we take into account Sirius' mysterious companion, astronomers do not know how the Sirius system could have been red less than two thousand years ago. Robert Gent, of the Astronomical Institute in Utrecht, Netherlands, has pointed out recently that there are in fact a multitude of ancient and classical references to the whiteness of Sirius, from the ninth century BC Persian Zend-Avesta, to the Chinese astronomer historian Sima Qian (*c.* 91 BC), the Roman writers Hyginus (*c.* 10 BC), Manilius (*c.* AD 15) and Avienius (*c.* AD 360), and the Spanish encyclopaedist Isidorus of Seville (*c.* AD 610).

It seems that Ptolemy and other classical writers must have simply made a mistake, perhaps confused by the fact that Sirius can appear a variety of colours from yellow to red when it is near the earth's horizon, in much the same way as the sun changes colour at sunrise and sunset, due to the absorbing and scattering effect of dust and molecules in the earth's atmosphere.

SIRIUS B, DARK STAR

Sirius' companion was the first example of 'dark matter' to be discovered. As I mentioned in chapter 2, Sirius was one of the three stars found by Edmund Halley to have changed their position significantly since the time of Hipparcos. Over a period of ten years from 1834, the German astronomer Friedrich Bessel studied this motion closely and found that it followed a wavy line. He concluded that Sirius was orbiting around an unseen companion. He deduced that the companion could not be a planet but must be of a mass comparable with Sirius itself and he estimated the orbital period as about fifty years. It took nearly twenty years to locate this companion and when it was found, by the American telescope-maker Alvan G. Clark in 1862, it turned out to be 10,000 times fainter than Sirius itself. Clark had just completed a new refractor which was at that time the largest in the world, with an 18½ inch diameter objective lens. He turned the telescope to Sirius and was delighted to find a very faint spot of light in just the position predicted by Bessel, a little bit further from Sirius than Uranus is from the sun. It is an extremely difficult object to study with a telescope, lost in the scorching light of the Dog Star, and it was another fifty years before Walter Adams, at Mount Wilson Observatory in California, was able to measure the colour of 'Sirius B' in 1915 and hence determine its temperature as about 9000 degrees centigrade. The very low luminosity of the star then implied that the star must be absolutely tiny, only about three times as big as the earth. The name *white dwarf* was coined for this type of star, of which many other examples are now known. In spite of its tiny size, the mass of Sirius B is about the same as the sun and so its mean density has the incredible value of about 100,000 times that of water. A matchboxful would weigh a ton!

White dwarfs are dead stars and nuclear reactions have ceased. The matter is so tightly packed together that all atomic structure has been destroyed and the electrons and atomic nuclei which make up atoms are crushed together. Normally an atom is like a solar system, with vast empty spaces between the electron orbits. In a white dwarf the matter is so tightly packed that the electrons are forced to squeeze past each other, in the process creating pressure which holds up the star against gravity. We will see in chapter 7 that to become a white dwarf is the eventual fate of stars like the sun. During the past twenty years even more dramatic collapsed objects have been found. *Neutron stars*, which are the relics left behind when massive stars explode as supernovae, came to light through the discovery of pulsars (see chapter 12). And the probable discovery of *black holes* has been one of the most exciting aspects of the development of X-ray astronomy (chapter 10). But the white dwarf Sirius B already began to give Bessel a headache 150 years ago.

Although these compact objects are all dead relics, they can cause dramatic fireworks when they are located in a close binary system, so that a companion star starts to dump material onto its compact neighbour. White dwarfs, for example, are now known to be responsible for two spectacular phenomena of this type, *novae* (chapter 11) and *Type I supernovae* (chapter 12). It is strange to think of this potential violence orbiting around the brilliant Sirius, the soul of the goddess Isis, the dog-star of evil omen.

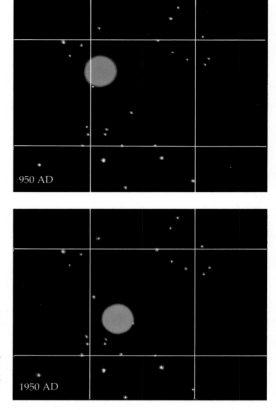

△ *Fig 3.4 The motion of Sirius on the sky over a thousand years. It moved about a third of a degree during this time*

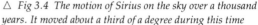

It were all one
That I should love a bright particular star . . .

SHAKESPEARE *All's Well That Ends Well*

Yes, these are the dog-days, Fortunatus . . .

W. H. AUDEN *'Under Sirius'*

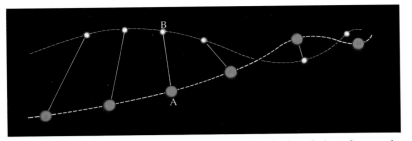

△ Fig 3.5 *The detailed path of Sirius A across the sky, showing how the irregular curved path is caused by the attraction of the faint white dwarf companion, Sirius B*

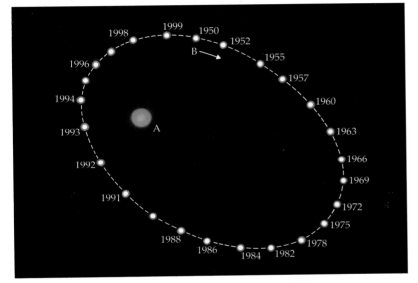

△ Fig 3.6 *The orbit of Sirius B about Sirius A over a 50 year period*

THE DOGON OF MALI AND SIRIUS B

In 1975 Robert Temple wrote a controversial book in which he described how, when in the 1930s the French anthropologist Marcel Griaule encountered the Dogon tribe of Mali in W. Africa, who had no advanced technology and in particular no telescopes, he found that the Dogon spoke of Sirius as having a dark, very dense companion. Since the Dogon could not possibly have known of Sirius B, Temple drew the extraordinary conclusion that the Dogon had been visited by intelligent beings from another planet.

However a simpler explanation (pointed out by Carl Sagan) would seem to have been that the Dogon were visited by other intelligent or at least well-informed beings from planet earth, for example missionaries, some time between 1915, when Adam's discovery was blazed across the world's newspapers, and Griaule's arrival in the 1930s. So far from having been visited by intelligent beings from other worlds, we seem to be horribly alone in this vast universe.

△ Fig 3.8 *... the star whose name men call Orion's Dog ...*

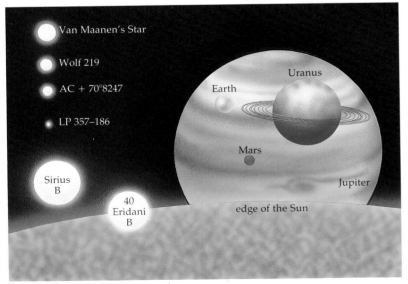

△ Fig 3.7 *Comparison of relative sizes of sun, planets and selected white dwarf stars*

C H A P T E R   4

To persons standing alone on a hill during a clear
midnight such as this, the roll of the world eastward
is almost a palpable movement.

THOMAS HARDY *Far From the Madding Crowd*

POLARIS

THE POLE STAR

Alpha Ursa Minoris, the brightest star in the constellation of the Little Bear, known as the North Star or Pole Star, is very close to the north pole of the heavens. Each night the constellations of the northern sky revolve about it in their slow dance. It was not until the sixteenth century, with the work of Nicolas Copernicus of Torun, that the cause of this rotation of the constellations about the Pole Star was firmly established as due to the rotation of the earth. Before that time many thinkers had suggested the idea, but it had never been universally accepted and the Platonic notion of a static, centrally placed earth remained the official view.

We do not feel the earth to be rotating. On a roundabout we feel the effect of centrifugal force thrusting everything away from the central axis, our perceptions are distorted by dizziness and we experience the rotation directly through the balance mechanism in our inner ear. Because we do not notice these effects when we stare up at the night sky, we have a strong subjective feeling that the earth is stationary. Though on nights when I have stared at the constellations for long enough to notice their motion, for several hours say, I have found that I experience something very like giddiness at the thought of our rotational speed of a thousand miles an hour. (Somehow the ten miles per second at which the earth moves through space in its orbit around the sun seems much more abstract.) The centrifugal force associated with the earth's rotation was detected in the seventeenth century, through the slightly flattened shape of the earth and the fact that when a weight hangs freely on a string, the string is not in fact quite vertical. Another effect of the earth's rotation, which we do not notice on a roundabout, is the Coriolis force, discovered by the Italian physicist Corioli in the eighteenth century. This was familiar to the artillery-men of the nineteenth century since it has the effect that a moving projectile swerves slightly to the right in the northern hemisphere and to the left in

△ *Fig 4.1 Close-up of Polaris*

the south. Other things being equal (which they are usually not), the bath-water swirls down the plug-hole in a clockwise direction north of the equator and in an anti-clockwise direction to the south.

The Pole Star is about one degree from the true pole of the heavens. Long-integration exposures, which trace out the rotation of the stars, show that Polaris itself moves in a small circle. But although Shakespeare has Julius Caesar say:

> . . . I am constant as the northern star,
> Of whose true fixed and resting quality
> There is no fellow in the firmament.
> The skies are painted with unnumb'red sparks,
> They are all fire and everyone doth shine;
> But there's but one in all doth hold his place . . .

the pole will not always stay near Polaris, nor has Polaris always been the nearest bright star to the pole. At the moment the pole is moving closer to Polaris and will reach its closest point, 0.46 degrees away, in the year 2102. Thereafter it will head into the constellation of Cepheus, passing close to the west of Casseiopea around the year 7000. The change is due to gradual change in the direction of the earth's axis in space. This was one of the greatest discoveries of antiquity, the *precession of the equinoxes*, made by Aristarcos of Samos in the third century BC. The true pole moves on a large circle 47 degrees in diameter every 25,800 years, or one degree every 175 years. Thus in the time of Julius Caesar, the pole was over 10 degrees away from Polaris and was equally near to Beta Ursa Minoris, the brightest of the stars which makes up the characteristic rectangle of the Little Bear. During the age of the pyramid builders of ancient Egypt, 4600 years ago, the star Thuban (Alpha Draconis) was the Pole star. And 15,000 years ago, Vega (Alpha Lyrae) was near the pole. To define north accurately enough for the traveller or seafarer of past millenia, however, it was only necessary to have a bright and recognizable star somewhere near the pole, so different millenia will have used different pole stars, each one being adequate for over a thousand years.

It is interesting to consider what the ancient seafarers of the southern hemisphere used for navigation. At present there are no bright stars near the celestial south pole. But three thousand years ago the south pole would have been close to the Small Magellanic Cloud (chapter 15), a prominent fuzzy patch of light in the southern sky, and this must have been a valuable signpost for the Polynesian navigators.

The concept of the pole star has been a resonant one throughout literature. Marlowe has Mephistopheles tell Faustus that the heavens:

> . . . all jointly move upon one axletree,
> Whose termine is termed the world's wide pole.

John Keats addresses his last sonnet to the Pole Star:

> Bright star, would I were steadfast as thou art –
>   Not in lone splendour hung aloft the night . . .

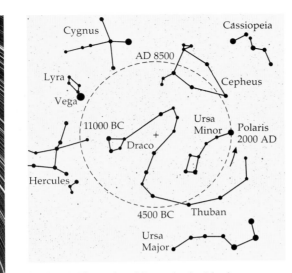

△ Fig 4.3 The motion of the north celestial pole over 26,000 years.

△ Fig 4.4 The trails of north polar stars. The bright arc near to the centre is due to Polaris itself

◁ Fig 4.2 The trails of south polar stars over the Anglo-Australian Telescope, during a long exposure lasting several hours

And Wordsworth muses on the use of the Pole Star for navigation:

> Chaldean shepherds, ranging trackless fields,
> Beneath the conclave of unclouded skies,
> Spread like a sea, in boundless solitude,
> Looked on the polar star, as on a guide
> And guardian of their course, that never closed
> His steadfast eye. . . .

For many centuries Polaris has been the Lodestar or Steering Star of navigators. Anglo-Saxon tribes, a thousand years ago or more, called it Scip-Steorra, or Ship-Star. Arab astronomers knew it as Al Kaukab al Shamaliyy, the Star of the North, or Al Kiblah, referring to its use to determine the direction of Mecca. The alleged steadfastness of the Pole Star is used by Confucius in the fifth century BC: 'He who rules by moral force is like the Pole star, which remains in its place while all the lesser stars do homage to it . . .' To the Pawnee Indians of Nebraska, Polaris is 'The Star That Does Not Walk Around'.

In legends found throughout Asia, Polaris is regarded as the spire or pinnacle of the cosmic Mountain of the World or Axis of the Universe, the fabled Mt Meru or Sumeru of Hindu or Buddhist lore. On the summit of Meru, 84,000 leagues above the Earth, dwell Indra, Surya, Vishnu and other deities. Many of the ancient temples of India and the Far East are symbolic representations of the cosmic mountain. In the texts of ancient China, the concept of the cosmic mountain appears as T'ai Chi, the Axis of the Universe, the Centre, the Ridgepole, the Great Root, the Wheel of the Universe, the Pole Star or the Wheel of Life. The North Pole of the sky was the home of Huan-T'ien Shang-Ti, the Supreme Lord of the Dark Heavens or the Supreme Prince of the North Pole. The cosmic mountain was identified with various actual peaks, usually with the T'ai Shan in eastern Shantung, sacred since at least the seventh century BC. Later the cosmic mountain gradually shifted in popular legend to the legendary peak of Khun-Lun or Tien Shan, immeasurably far to the west. In Taoist legends it is Shou Shan, the Jade Mountain of Immortality, where those sages who have attained ultimate wisdom dwell on the shores of the Green Jade Lake. More highly valued than gold, jade was revered for its unique power of putting a person in contact with cosmic forces and 'opening the doors of wisdom' to higher states of consciousness. Jade artefacts known as *hsun-chi*, consisting of a flat round disc with a circular hole and irregularly spaced notches on the outer rim, may have been used to locate the true celestial pole.

As a star, Polaris is a blue-green star about 360 light years away. It is about 1600 times more luminous than the sun. Besides not being a 'stead-fast' locator of the celestial pole, it is not steady in its light output either, being a variable star of the Cepheid type (see chapter 9), with a period of variability of just under four days. It has a fainter companion two thousand times the sun-earth distance away and may have a third, unseen close companion.

THE ZODIAC AND THE EQUINOXES

The celestial poles are the directions in the sky of the earth's axis of rotation. The *celestial equator* is the directions in the sky defined by the plane of the earth's equator. Astronomers then use a system of co-ordinates for stars analagous to longitude and latitude, but projected onto the sky. The celestial longitude is called *Right Ascension* and for historical reasons is measured in hours. (The Right Ascension measures the time at which the star is at its highest point, or zenith, at the time of the Spring Equinox, as viewed from the Greenwich meridian.) The celestial latitude is called the *declination* and is measured in degrees. Because of the Precession of the Equinoxes you also have to specify what year or 'Epoch' the co-ordinates refer to. At present astronomers use 1950 as the reference epoch, though our professional body, the International Astronomical Union, has decreed that we should be using the year 2000 now, an instruction which most of us are not taking a blind bit of notice of because it means changing the habits of a lifetime (pure laziness!).

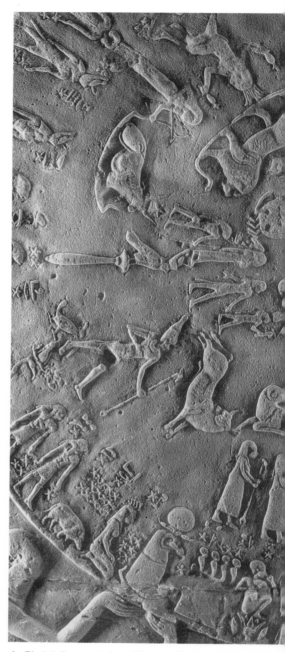

△ *Fig 4.6 Representation of the constellations of the zodiac from an ancient Egyptian temple ceiling at Dendera*

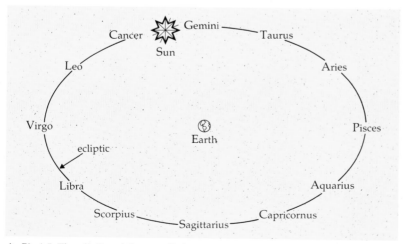

△ *Fig 4.5 The ecliptic and the constellations of the zodiac*

O, no! it is an ever-fixed mark,
That looks on tempests and is never shaken;
It is the star to every wandering bark,
Whose worth's unknown, although his height be taken.

SHAKESPEARE *Sonnet 116*

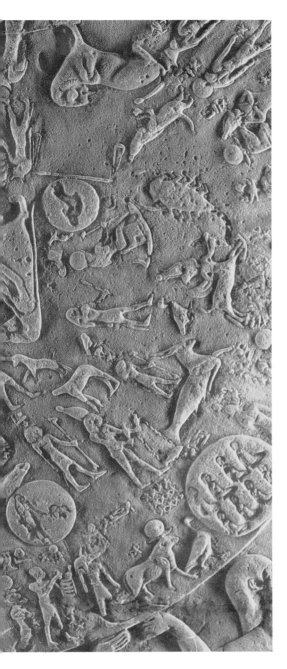

Another very important circle on the sky is the *ecliptic*, which is the path on the sky of the sun. Because the earth moves round the sun in a fixed plane, the sun appears to move round the earth also in a fixed plane, the ecliptic plane. The constellations which the sun passes through in its yearly path round the ecliptic are of especial interest and they have been called the constellations of the *zodiac*. The twelve constellations are Aries, Pisces, Aquarius, Capricornus, Sagittarius, Scorpius, Libra, Virgo, Leo, Cancer, Gemini, and Taurus (the Ram, the Fish, the Water-Carrier, the Goat, the Archer, the Scorpion, the Scales, the Virgin, the Lion, the Crab, the Twins and the Bull), and the sun spends a month in each one.

The *equinoxes* are the two points where the celestial equator meets the ecliptic. When the sun is at one of these points it lies directly over the equator at midday and the day and night are of exactly equal length. At present this happens on 21 March and 23 September. The spring equinox lay at the 'first point in Aries' in classical times, but the *precession of the equinoxes* moves the location of the celestial pole, the celestial equator, and hence of the equinoxes, and today the spring equinox lies in the constellation of Pisces. The precession of the equinoxes is caused by the action of the sun and moon's gravity on the earth's equatorial bulge.

Because the planets' orbits lie close to the plane of the earth's orbit, they also move along the ecliptic with time and so are to be found in the constellations of the zodiac. The irregular motions of the planets along the zodiac were a matter of great mystery to the astronomers and philosophers of ancient times. In many cultures – for example Babylonian, Egyptian, Greek, Chinese, Mayan and Aztec – these motions were associated with human destiny. Until the time of Copernicus, astrological prediction was the main reason that astronomers were employed by rulers and governments. The mystery of the motion of the planets was solved by the Copernican picture of the solar system. Astrology has continued its irrational existence to the present day. A measure of the credulity of the present age is that most bookshops in the Western world carry more books on astrology than on astronomy.

Oh if only once the sting of the air and the heat
of summer could make me hear
beyond sleep and death
the earth's axis, the earth's axis.

OSIP MANDELSTAM *Poems of the Thirties*

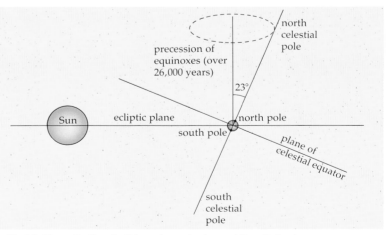

△ *Fig 4.7 The earth, the celestial equator and poles, and the ecliptic plane*

THE PLOUGH, OR BIG DIPPER

The Plough or Big Dipper is part of the constellation of Ursa Major, the Great Bear, and is one of the most easily recognized groups of stars on the sky. It is therefore especially convenient that the last two stars of the Plough point to Polaris, our millennium's pole star.

The constellation has been identified as a Bear not only in the Greek and Roman worlds, but also by the Iroquois, Cherokee, Blackfoot and Algonquin tribes of North American Indians, though the resemblance to a bear is not very striking. In Greek legend the bear was originally Callisto of Arcadia, daughter of king Lycaon. She was transformed into the animal to disguise her from Zeus's wife, the jealous and angry Hera. Callisto's young son Arcas, while out hunting, was about to kill the bear, not recognizing his mother in her strangely altered form. The gods then placed them both in the heavens as the Great and Little Bears. Ovid tells us in his *Metamorphoses* how Zeus

> . . . flung them through the air,
> In whirlwinds to the high heavens, and fix'd them there,
> Where the new constellations nightly rise,
> Lustrous in the northern skies.

It was because of the enmity of Hera that the Bears are not allowed to go to their rest beneath the rim of the earth like other constellations, but must circle the pole eternally.

Blind Milton found solace in imagining himself as a star-gazer:

> Let my lamp at midnight hour
> Be seen in some high lonely tower
> Where I may oft outwatch the Bear.

△ *Fig 4.8 The Plough, or Big Dipper*

△ *Fig 4.9 The constellation of Ursa Major, the Great Bear, acc* *he portrays the constellations as they would appear on a star-glo*

How swiftly the morning wind sweeps away!
The Great Bear glitters and curves down the sky.

LI HO (AD 791–817)

When the Wain of the first heaven
which setting nor rising never knew,
nor veil of other mist than of sun,

and which made there each aware of his duty,
even as the lower Wain guides him
who turns the helm to come into port

had stopped still . . .

DANTE *Purgatorio*

17th century Swiss astronomer Hevelius. Note that
front compared to the sky

My father compounded with my mother under the dragon's
tail, and my nativity was under Ursa Major, so that it
follows I am rough and lecherous. 'Sfoot! I should have
been that I am had the maidenliest star in the firmament
twinkled on my bastardizing.

SHAKESPEARE *King Lear*

I have been a hazel-tree, and they hung
The Pilot Star and the Crooked Plough
Among my leaves in times out of mind:

W. B. YEATS *'He thinks of his past greatness when a part
of the constellations of heaven'*

In *Othello*, Shakespeare conjures up the ferocious sea-storm which de-
stroys the Turkish fleet by imagining waves of such height that

> The wind-shak'd surge, with high and monstrous mane
> Seems to cast water on the burning Bear,
> And quench the guards of th' ever fixed pole . . .

In ancient China the Tseih Sing or Seven Stars were associated with the
celestial palace of the Lord on High and the heavenly mountain of Tien
Shan. The three stars of the Plough's handle are referred to in some
Chinese writings as the Jade Scales:

> The brilliant moon shines splendid in the night;
> One may hear the house cricket singing on the east wall . . .
> The Jade Scales speak of the beginning of winter,
> Scattering a million stars across the sky!

Ancient Britons saw the constellation as Arthur's Chariot, the Vikings as
the Chariot of Wotan or Odin. In Renaissance England the constellation
was known as Charles's Wain, shortened from Charlemagne's Wain (or
Wagon), and was popularly used as a natural clock. In Shakespeare's
Henry IV one of the porters at the inn-yard exclaims:

> Heigh-ho! An't be not four by the day,
> I'll be hanged; Charles' Wain is over the new chimney
> And yet our horse not pack'd . . .

Edmund Spenser, in the *Faerie Queen*, writes both of Polaris and Ursa
Major, compressing into a few brilliant lines the relationship of the Plough
and the Pole Star, the fact that the Plough never sets, the steadfastness of
the Pole Star and its use for navigation:

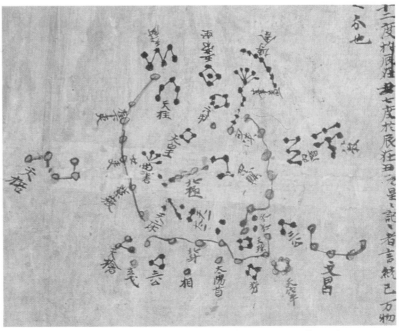

△ *Fig 4.10 Ancient Chinese star map with the Plough or Big Dipper (bottom)*

By this the northern wagoner had set
His sevenfold team behind the steadfast starre
That was in ocean waves never yet wet,
But firme is fixt, and sendith light from farre
To all that in the wide deep wandering arre . . .

And as late as the nineteenth century, Tennyson writes

We danced about the May-pole and in the hazel copse
Till Charles's Wain came out above
The tall white chimney tops . . .

If you look closely at the stars of the Plough, you will notice something interesting about one of the stars. The second star along from the end of the handle of the Plough, Mizar, is actually not one star but two. The companion was named Alcor (originally Al-jat, a rider) by the Arab astronomers of the Middle Ages. Curiously, they regarded this pair as a test of keen sight, though today they are easy to recognize. In 1651, Johannes Baptista Riccioli, who was professor of astronomy (among other subjects) at the University of Bologna, discovered that through the telescope Mizar itself becomes resolved into a close pair of stars, the first double star to be discovered with a telescope. So there are in fact three stars here, the double star system Mizar and its close neighbour Alcor.

The proper motions (p. 18) of the stars across the sky gradually distort the shapes of the constellations. The figure shows the Plough as it would have looked 100,000 years ago and as it will look 100,000 years from now. Five of the stars move together in space and make up part of the Ursa Major *moving cluster*, discovered by R. A. Proctor in 1869. It is the nearest such cluster of stars to the sun, at a distance of about seventy-five light years. It is interesting that when we look up at the northern sky, our eye is drawn so strongly to this very significant group of stars.

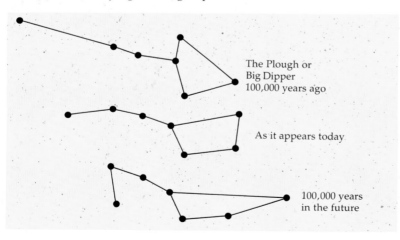

The Plough or
Big Dipper
100,000 years ago

As it appears today

100,000 years
in the future

◁ *Fig 4.11 How the appearance of the Plough/ Big Dipper will change over 200,000 years*

▷ *Fig 4.12 Van Gogh's 'Starry Night on the Rhone'*

C H A P T E R   5

Out on the lawn I lie in bed,
Vega conspicuous overhead
In the windless nights of June . . .

W. H. AUDEN *'A Summer Night'*

VEGA

A PLANETARY SYSTEM IN THE MAKING?

Alpha Lyrae, the Harp Star, is the fifth brightest star in the sky, dominating the summer sky in the northern hemisphere. The name Vega is derived from the Arabic Al Nasr al Waki, the Swooping Eagle; the alternative forms of Waghi, Vagieh and Veka also appear on medieval charts, where the star and its constellation are depicted as an eagle, vulture or falcon, often bearing a harp or lyre in its beak or talons. The Greek name was kithara, the harp or lyre of the ancient Greek bards, of the god Apollo and of Orpheus.

In ancient China, Vega figures in the legend of the Herd-Boy and the Weaving-Girl, which is mentioned in the *Shih Ching* or *Book of Songs* of the sixth century BC. The latter is, incidentally, one of the books which the

▷ *Fig 5.1 The constellation of Lyra, dominated by Vega (below centre)*

megalomaniac emperor Shih Huang Ti, builder of the Great Wall, ordered to be burned in the third century BC. In the legend, Vega is the Weaving-Girl or the Lady of the Han River, while the Herd-Boy is Altair and the two stars on either side of it. The young lovers, absorbed in each other, neglected their duties to the gods and were punished by being separated by the Celestial River, the impassable barrier of the Milky Way. Once a year, however, on the seventh night of the seventh moon, the lovers are allowed to meet when a bridge of birds temporarily spans the River of Stars.

The Pawnee Indians of Nebraska call Vega the Black Star which, as we will see below, is rather an appropriate name.

At northern latitudes Vega lies directly overhead in the evening hours of late July and August. It has a sharp blue-white glitter, which in a large telescope becomes a vivid blue pool of light. It is twenty-seven light years away from us, fifty-eight times as luminous as the sun and three times more massive, a *blue giant* star. Its diameter has been measured with a special telescope, at Narrabri in Australia, which overcomes the scintillations of the earth's atmosphere by making the rays from different parts of the star interfere together to make a fringe pattern on the detector. The scintillations are what makes stars twinkle and mean that although distant stars are highly point-like, they can never produce an image smaller than about 0.5 arcseconds (1 7000th of a degree) with a normal telescope. The naked human eye, however, can only resolve images which are at least one arcminute in diameter (1 60th of a degree) and so the stars never look smaller than that to the naked-eye observer. The Narrabri measurements showed that the diameter of Vega is about 3.2 times that of the sun. Vega was in fact the first star ever to be photographed, in 1850 at Harvard Observatory, using the daguerreotype process with an exposure of a hundred seconds.

Infinitely apart lie the Herd-Boy star
And the streaming whiteness
Of the Lady of the Han River,
Working endlessly at her loom . . .
. . . How vast a distance separates them!
Always the immeasurable River yawns before them . . .

TU FU AD 713–770

△ Fig 5.3 Launch of the IRAS Infrared Astronomical Satellite

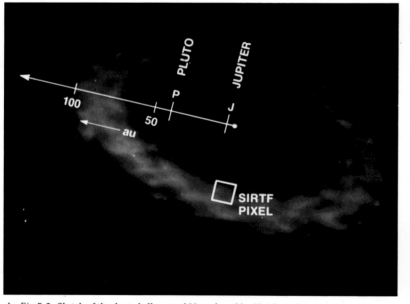

△ Fig 5.2 Sketch of the dust shell around Vega found by IRAS, with scale of the solar system and resolution of a planned space mission, SIRTF, overlaid

Since the last time Mr Palomar looked at the stars weeks or months have gone by; the sky is all changed. The Great Bear (it is August) is stretched out, almost lying down, on the crowns of the trees to the north-west; Arcturus plunges towards the outline of the hill, dragging all the kite of the dipper with him; exactly west is Vega, high and solitary . . .

ITALO CALVINO *Mr Palomar*

In 1983 the Infrared Astronomical Satellite, IRAS, made a remarkable discovery about Vega. Because the star is very bright and extremely well-studied at optical wavelengths, it had been chosen as a calibration standard for the IRAS survey. We had assumed that we could simply extrapolate the visible and near infrared measurements to the far infrared wavelengths studied by IRAS. However George Aummann and Fred Gillett, two American members of the IRAS team working on the calibration problem at the IRAS ground-station at the Rutherford-Appleton Laboratory in Oxfordshire, England, found that there was something very unusual about Vega. While the other bright calibration stars behaved as predicted, Vega was much brighter than expected in the far infrared. The excess emission appeared to be coming from small particles of dust at a temperature of −193 °C, compared with the temperature of the star's surface of 10,000 °C. The inference was that there is a thin screen of solid particles ('dust') situated at about eighty times the sun-earth distance from Vega. The observations also allowed the deduction that the particles are relatively large compared with most samples of interstellar dust, at least one millimetre in diameter.

Now Vega is known to be a very stable star which does not vary its light output or eject any material from its surface. The dust shell surrounding Vega must therefore have been there since the star was formed. The amount of material involved turns out to be comparable to that involved in the planets of the solar system. We seem to be seeing a planetary system in the making.

The IRAS scientists immediately checked out other stars similar to Vega and found some other examples with dust shells, of which the best studied case is Beta Pictoris. Ground-based observations by Bradford Smith and Richard Terrile at the Las Campanas Observatory in Chile have shown that the dust round this star is distributed in the form of a disc, just as would be

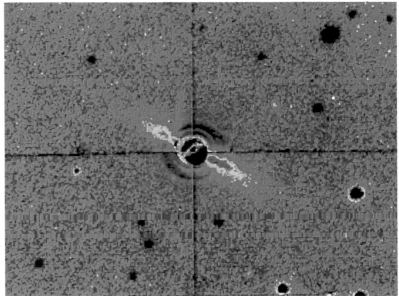

△ *Fig 5.4 False colour picture of the dust shell around Beta Pictoris, as seen in scattered light*

expected for a proto-planetary disc. They found that the disc has an overall diameter of 800 sun-earth distances but that the inner part of the disc, within 30 sun-earth distances of Beta Pictoris, seems to have been cleared of dust. This zone is comparable in size to the region occupied by the planets of the solar system, so perhaps the material within that radius has already formed into planets. The existence of this dust-free zone is still a matter of controversy, however, and there is certainly no direct evidence for planets round Vega or Beta Pictoris.

The search for planets round other stars has been a long and so far fruitless one. There are no reliable cases of planets even of Jupiter size, and the detection of a planet like the earth is beyond the reach of present instrumentation. The best hope lies in studies of the motion of stars in the sky like those which led to the discovery of Sirius B in the nineteenth century. To date, several examples of companions about five times the mass of Jupiter have been found, though not all have stood up to close scrutiny. Perhaps the Hubble Space Telescope, launched on 24 April 1990, will have better success. Why is it so important for us to find other planets? Because we know that no matter how majestic and powerful the universe and its phenomena may be, our planet and its life are the most significant parts of that universe we have yet seen. We need to know whether earth and its life are unique in the universe: so far, we have no evidence that it is not.

One final interesting fact about Vega is that the sun and other very nearby stars are moving through space in a direction rather close to that of Vega. William Herschel, in 1783, was the first to measure this effect, using the known proper motions of just twelve stars. The basic idea is that because the sun and its planets are, in the words of Buckminster Fuller, 'flying formation' through space, we seem to see the nearby stars coming towards us with, on average, the same speed in the opposite direction. Modern studies, based on the proper motions of tens of thousands of stars, come up with an answer not very different from Herschel's. The sun is travelling through space at twenty kilometres per second and in only 400,000 years we will arrive near Vega – or rather, where Vega is now, because naturally Vega too has its own motion. The stars that seem so static and permanent are weaving about in a complicated dance.

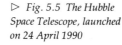

▷ *Fig. 5.5 The Hubble Space Telescope, launched on 24 April 1990*

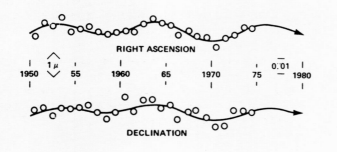

RIGHT ASCENSION

| 1 μ | | | | 0ʺ01 | |
| 1950 | 55 | 1960 | 65 | 1970 | 75 | 1980 |

DECLINATION

◁ *Fig 5.6 The small changes in position of Barnard's star with time, which Peter van de Kamp of Sproul Observatory believes are due to two companions of mass 0·4 and 1·0 times that of Jupiter. Upper curve: changes in right ascension, lower curve: changes in declination*

THE SEARCH FOR PLANET X

In addition to our natural curiosity about other planetary systems, much interest has focused on whether there could be a tenth planet in our solar system. Only the five naked eye planets were known to the ancients. William Herschel discovered Uranus in 1781 during his telescopic survey of the northern sky, Neptune was discovered in 1846 from its perturbations to Uranus's orbit and Pluto was found by Clyde Tombaugh in 1930 during a programme designed to discover the cause of small unexplained perturbations to the orbit of Uranus and Neptune. However we now know Pluto is too small to have caused these perturbations, which remain unexplained. The discrepancy is even more acute if observations of Neptune by Galileo in 1613 and by Lelende in 1795 were correctly recorded. Galileo was observing Jupiter at a time when Neptune was nearby and appears to have recorded the position of Neptune, but not quite where it should have been if there were no tenth planet. Lelende was measuring the positions of stars and recorded a 'star' where no star should have been. In retrospect it is calculated that Neptune should have been close to the position recorded by Lelende, but again the position is not quite correct.

On the other hand the tracking of the Pioneer and Voyager spacecraft on their voyages to the outer planets of the solar system and beyond show that there can be no tenth planet in the plane of the ecliptic today. The orbit would have to be highly tilted to that of the orbit of the earth and other planets. Several astronomers still believe that Planet X does exist and are continuing to search for it.

There were two occasions in 1984 when I thought I had found Planet X in the data from the IRAS Infrared Astronomical Satellite. The first was when a bright cool source turned up on the IRAS map of the Galactic Centre region, lying right on the plane of the ecliptic. It soon turned out that the source was not moving across the sky, as a planet would, and its infrared spectrum showed that it was in fact a heavily dust-shrouded red giant star. However rumour somehow reached the magazine *New Scientist* and nothing could persuade them not to print the story 'IRAS discovers tenth planet'!

On the second occasion I was carrying out a more systematic search for cool, moving objects when to my amazement one turned up. Unfortunately this turned out to be Comet Bowell, which had been discovered a year or so previously out beyond Jupiter. It is still not completely impossible that Planet X lies hidden somewhere in the IRAS data.

C H A P T E R 6

Nights, I squat in the cornucopia
Of your left ear, out of the wind,
Counting the red stars and those of plum-colour.

SYLVIA PLATH *'The Colossus'*

MIRA

THE FIRST KNOWN VARIABLE STAR

Omicron Ceti, Mira, the Wonderful, changes its brightness every six months by a factor of more than a hundred at visible wavelengths. This variation in brightness was first noticed by the Dutch astronomer David Fabricius on 13 August 1596, though he does not seem to have followed up this discovery. The star was recorded in Johann Bayer's Star Catalogue of 1603 as the fifteenth brightest in the constellation of Cetus, the Whale, so he gave it the fifteenth letter in the Greek alphabet, Omicron. Another Dutch astronomer, Johannes Holwarda, studying the star in 1638, noticed that it had been previously seen by Fabricius and Bayer and realized that it must be a star that varied its brightness dramatically. The Latin name, Mira, the Wonderful, was suggested by the one of the leading astronomers of the day, Hevelius of Danzig, in 1648.

▷ Fig 6.1 Parts of
Eridanus and Cetus
constellations, with Mira
(Omicron Ceti), close to its
maximum brightness, the
bright star to the upper
right

The period of variation is about 331 days and every maximum since 1638 has been observed. Monitoring the changes in variable stars is one of the activities carried out today by dedicated amateur astronomers. Astronomy is probably the last science in which amateurs still play a role. Grote Reber kept radio astronomy alive during the 1940s after its discovery by Karl Jansky in 1932. He used to come home from his job with a radio valve company and spend his evenings working with the radio dish which mystified his neighbours. It was very different in earlier centuries when most scientists were amateurs with independent means. In seventeenth-century England, the only professional astronomers were the three university professors of astronomy at Cambridge, Oxford and Gresham College, London. Flamsteed was appointed as Astronomer Royal at Greenwich, it is true, but his income derived from an ecclesiastical living, the work for which was carried out by lowly paid curates. At the end of the eighteenth century, an amateur of wide interests like Goethe could still make significant contributions to several branches of science.

Mira varies because its whole surface is pulsating in and out. The diameter of Mira at maximum is four hundred times that of the sun, and two hundred times at minimum, yet its mass is no more than twice that of the sun. It is a *red giant* star and is approaching the end of its life. At minimum light its luminosity at visible wavelengths is slightly less than that of the sun, while at maximum light it is on average 250 times more luminous than the sun. The dramatic variations at visible wavelengths are misleading, though. The total power at all wavelengths, including infrared wavelengths where Mira emits most of its energy, varies only by a factor of 2.5 from maximum to minimum light. It is the change in temperature, from 2200 to 1700 degrees centigrade, which plunges Mira into invisibility.

The dramatic pulsations of red giant stars are accompanied by a ceaseless flow of matter from their surfaces in a shrieking wind at twenty kilometres per second in all directions. When the matter in this wind has travelled far enough from the star and cooled, solid dust particles condense out. Depending on the composition of the star, the particles may be magnesium or iron silicates, similar in composition to the rocks and stones of the earth, or amorphous carbon particles similar to those in smoke. After a time the star may be hidden from view at visible wavelengths, shrouded in a cloud of dust of its own making. I spent some years studying the infrared emission from these *circumstellar dust shells*, and became fascinated by the intricacies of the flow of radiation through dust. This interest in the flow of radiation through dust and smoke was what drew me into a study of the Nuclear Winter effect, which I have written about in *Fire and Ice: the nuclear winter*.

A star like the sun hardly changes for thousands of millions of years. During this time the power-house at its centre is a thermonuclear one, fusing hydrogen atoms together to make helium in a process which humans have mimicked in the manufacture of hydrogen bombs. When the hydrogen in the hot core of a star like the sun is exhausted, the whole structure of the star will start to change. The centre contracts and heats up. Hydrogen continues to be fused to helium in a shell surrounding the helium core. While this is going on, the outer layers spread out enormously and cool down, so the star becomes a red giant. Thus Mira represents the future fate of the sun.

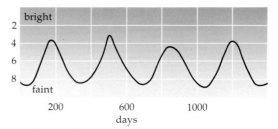

△ *Fig 6.2 The variations in the brightness of Mira with time over a four-year period*

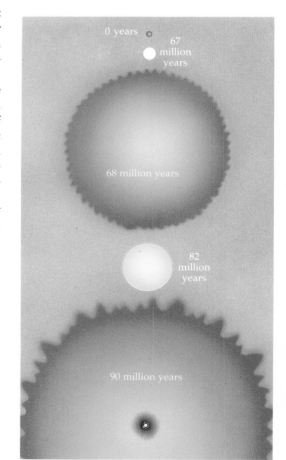

△ *Fig 6.3 Schematic illustration of the evolution of a star of five times the mass of the sun, after it exhausts the hydrogen in its core. At first the core of the star heats up while the envelope grows enormously in size to form a red giant. When the centre is hot enough for helium-burning to commence, the star undergoes a rapid change to a more compact, hotter state. Then as the helium is exhausted the star becomes a red supergiant. During this phase there is rapid mass-loss from the surface of the star, culminating in the ejection of the surface layers as a planetary nebula*

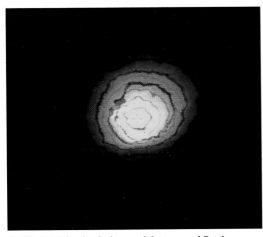

△ *Fig 6.4 The cloud of gas and dust around Betelgeuse seen in visible light*

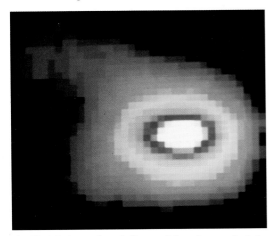

△ *Fig 6.5 A vast dust shell around the red supergiant star R Coronae Borealis as mapped in infrared light by the IRAS Infrared Astronomical Satellite. The shell, 25 light years across, was ejected from the star thousands of years ago*

I had a dream, which was not all a dream.
The bright sun was extinguish'd, and the stars
Did wander darkling in the eternal space,
Rayless, and pathless, and the icy earth
Swung blind and blackening in the moonless air . . .

BYRON *'Darkness'*

We are oppressed by oblivion,
by the idea of nothing at all.

SAPPHO *'Someone, I tell you' c.* 6 BC

Eventually the centre of the star becomes hot enough for the next fuel, helium, to start to be fused to carbon. The moment when helium-burning begins is a dramatic one because the core of the star undergoes a minor explosion. The whole star changes its structure in a matter of months and becomes a yellow giant, with a much hotter surface. At this stage nuclear reactions are taking place in two zones. Helium is being fused to carbon at the centre of the star and hydrogen is being fused to helium in a spherical shell a bit further out from the centre. As the helium in turn begins to be exhausted, the star grows larger and cooler, becoming a red giant once more, this time at even higher luminosity. When helium becomes exhausted at the centre of the star, helium-burning continues in a spherical shell surrounding the carbon core, so there are now two nuclear-burning shells. This is the stage that Mira is at now.

Mira has not long to live. It is losing mass profusely in the wind puffed off by its pulsations. Soon (within ten thousand years) a convulsion will overtake the star in which the whole outer layers are blown off in a vast smoke ring to become a 'planetary nebula' (chapter 7). The sun will follow the same route in some five thousand million years time. In *The Time Machine*, H. G. Wells portrays the dying sun as pallid and red, but in fact the huge increase in the sun's luminosity as it becomes a red giant will destroy all life on earth. The earth may well be engulfed in the sun's outer layers. Long before then, life on earth will either have ended or the great migration to planets around other nearby stars will have taken place. This impending death of the sun is not an event in our own lives, but it is a certain event in the history of human civilization. It is one of those facts about existence, like the finite age of life on earth and the finite age of the universe itself, that we just have to swallow.

Another very prominent variable star is Betelgeuse, Alpha Orionis, the brightest star in the constellation of Orion. The name, from the Arabic Beit Algueze or Ibt al Jauzah, is usually translated as Armpit of the Giant, the Giant in question being Orion the Hunter. The variations of Betelgeuse, which are rather irregular, were probably first noticed by John Herschel in 1836. In his *Outlines of Astronomy* published in 1849 he wrote: 'The variations of Alpha Orionis, which were most striking and unequivocal in the years 1836–1840, within the years since elapsed became much less conspicuous.' By December 1852, however, it had brightened so much that Herschel thought it the brightest star in the northern hemisphere. It also had very bright maxima in 1894, 1925, 1930, 1933, 1942 and 1947. The main period of variation is about 5.7 years but there may be shorter variation periods of 150 to 300 days superposed.

The red colour of Betelgeuse is apparent even under the rather poor observing conditions of walking home along a London street. It is not only a lot brighter than Mira – which is quite a difficult star to spot for the armchair astronomer, even at maximum light – Betelgeuse is also very much more luminous because it is considerably further away, at some 550 light years. At maximum light it is 14,000 times as luminous as the sun, and 7600 times as luminous at minimum light. The angular size of Betelgeuse has been measured using 'interference' methods (see p. 40) and is found to be about $\frac{1}{20}$th of an arcsecond, or 1 70,000th of a degree. The corresponding diameter is seven hundred times that of the sun. Such a star is called a *supergiant*.

The mass of Betelgeuse is about twenty times that of the sun, so it is using up its available energy about five hundred times faster than the sun and will live for only ten million years in all. When young, such a star would be a luminous blue supergiant. Betelgeuse is close to the end of its life, having exhausted the hydrogen in its core and become a cool red supergiant. It is now in the process of working its way through a sequence of nuclear fuels – helium, carbon, nitrogen, oxygen, and so on through magnesium, aluminium and silicon – until an energy crisis is reached when the core is composed of iron. Within the next ten thousand years Betelgeuse will blow up in an extremely dramatic and impressive supernova explosion, which will be visible in the daytime for many months.

Mira and Betelgeuse are prototypes of variable red giants and supergiants and many other examples are known. Eta Carinae, however, in the southern hemisphere, is a unique object. It was first recorded by Edmund Halley in 1677, in the course of his survey of the southern skies from the island of St Helena. It was visible to the naked eye but not especially bright. In subsequent years it was found to vary irregularly, becoming moderately bright in 1730 and then again in 1801 before fading for several years. Then in 1820 the star started to brighten steadily. In April 1843 it had become the second brightest star in the sky, only slightly fainter than Sirius. After that it began to decline steadily and ceased to be visible to the naked eye in 1868. What is the cause of this extraordinary behaviour?

There are a number of clues to the nature of Eta Carinae, though the riddle is by no means solved even today. Firstly it is located near the spectacular 'Keyhole Nebula', a brilliant and complex cloud of dust and gas being illuminated by newly formed massive stars. A distance of 3700 light years has been derived for the star, which means that at its 1843 maximum Eta Carinae was the most luminous star in the whole of our Galaxy, several million times as luminous as the sun. To explain this luminosity we must suppose that the star is one of the most massive known, about one hundred times as massive as the sun. Eta Carinae is surrounded by a wispy shell of gas which can be seen to be expanding outwards and is presumed to be associated with the 1843 outburst. These wisps are especially rich in nitrogen, which must have been made in nuclear reactions in the core of the star. This shows that the star is fairly far advanced along its evolutionary life and has not long to live. Finally, in the past twenty years it has become apparent that Eta Carinae is today emitting most of its energy at infrared wavelengths and that it is shrouded in a thick cloud of dust. The total luminosity at all wavelengths today is not very different from that at the 1843 peak. We seem to be seeing the last stages of evolution of a very massive star, one that is even closer to blowing up as a supernova than Betelgeuse. The denouement could come at any time.

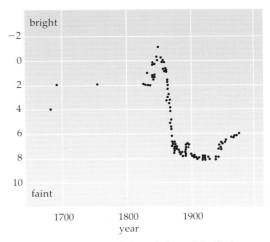

△ *Fig 6.7 The visible light variations of Eta Carinae over the past three centuries, with its spectacular peak in 1843*

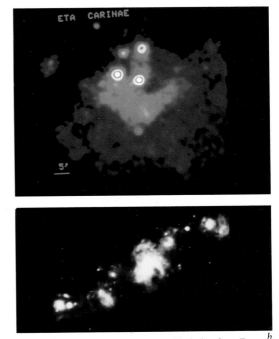

△ *Fig 6.8a Eta Carinae in X-rays. Emission from Eta Carinae (bright spot above centre) and several other hot stars is seen.*
b A larger-scale view of Eta Carinae (centre) in the infrared

I ofen looked up at the sky an' assed meself the question – what is the stars, what is the stars?

SEAN O'CASEY *Juno and the Paycock*

◁ *Fig 6.6 The remarkable variable star Eta Carinae (upper right) and the Keyhole nebula*

C H A P T E R   7

... these stars
whose light speaks a known language ...

BAUDELAIRE *'Obsession'*

RING NEBULA

DEATH THROES OF A
STAR LIKE THE SUN

The Ring Nebula was the first 'planetary nebula' to be discovered, by the French astronomer Antoine Darquier of Toulouse in 1779. He described the 'disc as large as Jupiter, but dull in light and looking like a faded planet'. The French astronomer Charles Messier, looking for the comet of 1779, found it a short time later: it is number 57 in his Catalogue of Nebulous Objects. Before the eighteenth century only a few astronomical objects had been recognized as distinct from stars in not being point-like. The Andromeda Nebula (chapter 16), the Magellanic Clouds (chapter 15) and the Orion Nebula (chapter 8) were the best known. The fuzzy, cloudy appearance of these objects led to the general term *nebulae* (Latin for clouds) for all extended objects. Messier, interested in finding new comets, made a list of the 103 most prominent examples of 'nebulous objects or clouds of stars', which could be confused with comets.

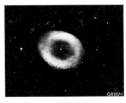

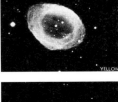

▷ *Fig 7.1 The Ring Nebula, a hollow shell of gas which has been blown off a dying star and is lit up by the hot blue star in the centre*

△ *Fig 7.2 Four views of the Ring Nebula in blue, green, yellow and red light*

Messier thought the Ring Nebula 'possibly composed of very small stars'. William Herschel began observing the 103 objects in Messier's list in October 1783 with his sister Caroline, using his new twenty-foot telescope. They found that most of Messier's nebulous objects turned out to be clusters of stars when seen through a large telescope, and they were at first confident that all would eventually be resolved into stars. In a paper published in June 1784, Herschel includes Messier 57 (or M57 for short) in a list of nine objects which 'shewed a mottled kind of nebulosity, which I shall call resolvable; so that I expect my present telescope will, perhaps, render the stars visible of which I suppose them to be composed.' Herschel was a bit wide of the mark here. The other eight objects included M1 (the Crab Nebula, a supernova remnant – see chapter 12), M27 (the Dumb-bell Nebula, another planetary nebula), and four galaxies (M33, M81, M82 and M101), which were not to be resolved into stars for another two hundred years. Only M3 and M79 were in fact star clusters.

Less than eight months later, in a paper published in February 1785, Herschel described M57 as a perforated nebula, or ring of stars: 'Among the curiosities of the heavens should be placed a nebula, that has a regular, concentric, dark spot in the middle, and is probably a Ring of stars.' In the same paper Herschel classified M27, the Dumb-bell Nebula, as a double star. He did now recognize a new class of nebulae, which he called planetary nebulae:

> I shall conclude this paper with an account of a few heavenly bodies, that from their singular appearance leave me almost in doubt where to class them. . . . The planetary appearance of the two first is so remarkable, that we can hardly suppose them to be nebulae; their light is so uniform, as well as vivid, the diameters so small and well defined, as to make it almost improbable they should belong to that species of bodies.

Thus Herschel recognized that 'planetary nebulae' were the fly in the ointment for the hypothesis that all nebulae were star clusters. Six years later in a paper entitled 'On Nebulous Stars, properly so called', published in February 1791, he found examples of planetary nebulae with central stars visible and concluded that 'the nebulosity about the star is not of a starry nature'. He believed that these were cases where the central star was being regenerated by the material in the nebulosity. By 1811 he was classifying most galaxies in a similar category, as gaseous objects condensing to become stars. Thus it was planetary nebulae which forced Herschel away from his essentially correct guess of 1784 that most of the nebulae he was seeing were very distant star systems. It was to be 140 years before it was finally established that most nebulae lie far beyond our Milky Way Galaxy.

▷ *Fig 7.3 Other examples of 'planetary' nebulae. They have no connection with planets, but William Herschel, who first classified this type of object, thought they looked like planets, and the name has stuck.*
a The 'Dumb-bell'
b NGC6543
c NGC6302, the 'Bug' Nebula
d The 'Helix'

a

c

b

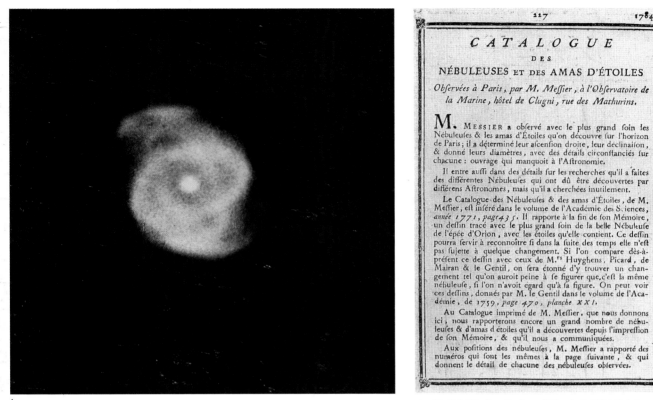

△ *Fig 7.4 The frontispiece of Messier's 1784 Catalogue of Nebulous Objects and Star Clusters*

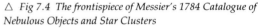

d

△ *Fig 7.5 William Herschel*

SPECTRAL LINES

The science of astronomy was transformed by the invention of the spectroscope in the nineteenth century. The image of a star in a telescope is made to fall on a narrow rectangular slit and then viewed through a prism. The prism splits the light from the star into all its constituent wavelengths, the *spectrum* of the star. Each colour, or wavelength, gives rise to an image of the slit, and these rectangular slices are laid out in a line, side by side.

Now, if the star is shining more or less equal amounts of every wavelength, then the resulting spectrum will merge into a smooth multicoloured swathe like the rainbow. If, however, the star is shining very brightly in certain specific wavelengths, for example the characteristic wavelengths of the element hydrogen, or those of the element iron, then bright lines will appear across the spectrum at these wavelengths, called *emission lines*. If, on the other hand, all the light of certain wavelengths has been absorbed by material in the surface layers of the star, then dark lines will appear across the spectrum at these wavelengths, called *absorption lines*. These spectral lines are a powerful diagnostic of the composition of the star, since we know from laboratory studies on earth what the characteristic wavelengths of all the most abundant elements are.

The element helium was first identified on the sun through its characteristic spectral lines and only later discovered on earth.

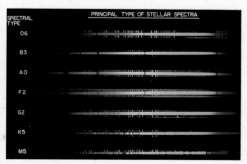

Examples of the main types of stellar spectra. The hottest star (type 0) is at the top and the coolest (type M) is at the bottom. Notice the bright emission lines in the spectra of the hot stars and the dark absorption lines and bands in the spectra of the cooler stars.

The name planetary nebula is highly misleading, because they have nothing whatever to do with planets. The ring which characterizes the Ring Nebula is in fact a luminescent shell of gas, rapidly expanding away from a central star, which is a bluish dwarf star with a temperature of 100,000 degrees centigrade. This star was first noticed by F. von Hahn in Germany in 1800 and is the exposed core of a star like the sun after its outer layers have been blown off in the final convulsion of its Mira phase (p. 47). Nuclear fuels are exhausted and the core has nothing more to do but cool off towards eventual death, first as a white dwarf like Sirius B and finally as a black dwarf.

The intense ultraviolet radiation from the very hot central star strips electrons off the atoms of the gaseous shell, leaving them *ionized*. Free electrons captured by such atoms may have far more energy than the outer electrons in an atom usually have. The electrons dispose of this excess energy by emitting the difference in energy as light of a particular wavelength, the phenomenon of *fluorescence*. These wavelengths show up as bright lines across the spectrum of the nebula when its light is passed through a prism spectrometer. The typical bluish-green colour of planetary nebulae, for example, is due to two bright emission lines of doubly-ionized oxygen (two electrons stripped off each atom) in the green part of the spectrum. These spectral lines were not understood until the 1920s and were attributed at first to an unknown element, 'nebulium'. The problem was that these lines are rarely generated under normal laboratory conditions. When they were finally understood, they were given the inaccurate name 'forbidden' lines.

The shells of planetary nebulae are found to be rich in carbon, nitrogen and oxygen, which demonstrates that these are stars which have progressed to the state of helium burning. The carbon, nitrogen and oxygen are by-products of the helium-burning process, dredged to the surface of the star by the process of convection, analogous to the turbulent motion of water in a boiling saucepan, and then blown off the star with the rest of the surface layers. To reach the stage of helium ignition, a star must be at least as massive as the sun. On the other hand, if a star is more massive than about eight times the mass of the sun, it will eventually proceed to ignite carbon and other fuels until an iron core is produced, at which stage it will blow up as a supernova (see chapter 12). Planetary nebulae are therefore the death throes of stars in the range 1–8 solar masses. As carbon, nitrogen and oxygen are the key elements in the production of life, the beautiful shells of planetary nebulae are of more than just aesthetic interest. These key elements of our bodies were fused in stars like the sun and thrown out into the interstellar gas in long-vanished planetary nebulae events thousands of millions of years ago, to be collected together into the cloud of gas and dust from which our solar system formed.

One of the problems not yet fully solved by astronomers is how the transition from a red giant star like Mira to a planetary nebula like the Ring Nebula takes place. The process is rather a rapid one in astronomical terms, taking at most ten thousand years. During the final stages of its life as a red giant, the star loses mass so profusely that it becomes completely shrouded in dust and virtually no visible light emerges from the star. The dust grains radiate in the infrared and this radiation can escape from the dust cloud, which is seen only as a strong infrared and microwave source.

Many such sources have been found by mapping the sky at the characteristic microwave wavelengths of the hydroxyl (OH) molecule. Because they are also very bright infrared sources, they are called OH–IR sources by astronomers. Like Mira these objects are still undergoing irregular pulsations, but with much longer periods of variation, often several years.

Planetary nebulae are very varied in their appearance and in their density structure. Some appear to be the result of a single convulsive ejection of a huge mass of gas, as much as the mass of the sun. Others have a more complicated structure, suggestive of several ejections or of an ejection continuing erratically over a prolonged period of time. The picture is complicated by the fact that the central stars of planetary nebulae are continuing to lose mass in a very high-velocity wind. Over a thousand planetary nebulae are known in our Milky Way galaxy, and the true number is probably at least ten times that. How a star like Mira becomes a planetary nebula like the Ring Nebula is one of the great unsolved problems of contemporary astronomy.

I believe a leaf of grass is no less than the journey-work of the stars.

WHITMAN *Song of Myself*

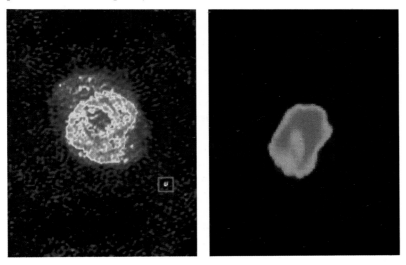

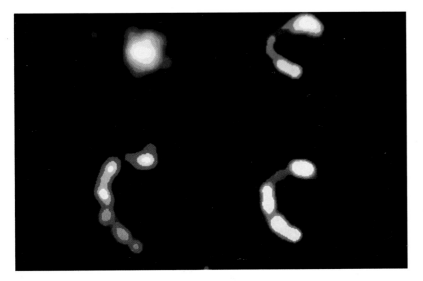

◁ *Fig 7.6 (left) Radio images of planetary nebulae*
a. NGC6543. The radio emission is from the hot gas.
b. NGC6302. The shell of gas is expanding at 10 kilometres per second and was ejected by the central star only 5000 years ago

▽ *Fig 7.7 (below left) Radio maps of the expanding shell of dust and gas around a red giant star which is an 'OH-IR' source. Radio emission from the hydroxyl (OH) molecule is mapped at four different frequencies, picking out the material at four different expansion velocities*

▽ *Fig 7.8 (below) The planetary nebula NGC7027 imaged at a characteristic infrared wavelength of molecular hydrogen. Two bright lobes of emission can be seen near the centre*

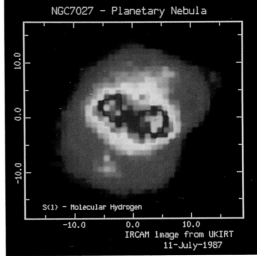

C H A P T E R 8

The small stars are trembling
Orion and the Pleiades.

ANON. *The Book of Songs*, 6th Century BC

ORION NEBULA

THE BIRTH OF NEW STARS

The brilliant constellation of Orion, the Hunter, is visible from every inhabited part of the earth. For northern hemisphere observers it dominates the southern winter sky: in the southern hemisphere it is a northern summer constellation. Orion is a somewhat shadowy figure. Homer refers to him as 'the tallest and most beautiful of men'. He had fallen in love with the divine huntress Artemis but was unintentionally killed by an arrow from her bow. In another version of the story, Orion was killed by the sting of the deadly scorpion sent by Hera to punish him for his arrogant pride. Nevertheless he was honoured by a place in the heavens and the fatal scorpion was placed in the exact opposite part of the sky so that it could never harm him again. Orion is the giant who is said to have pursued the Pleiades, the Seven Sisters, and was consequently blinded by the angry

▷ *Fig 8.1 The constellation of Orion (left), with the V of Taurus to the upper right*

△ *Fig 8.2 The constellation of Orion, the Hunter, according to the 17th century Swiss astronomer Hevelius*

king of Chios. On the advice of Hephaestos, however, Orion climbed to the top of a great mountain on or near the island of Lemnos, where as he faced the rising sun his sight was restored.

In classical times Orion was associated with winter storms. Polybius in the second century BC attributes the destruction of the Roman fleet during the First Punic War to the fact that it unwisely sailed with the rising of Orion. Virgil, Pliny and Horace refer to Orion as 'the bringer of clouds', 'the stormy one' or 'he who brings peril on the seas'. In the seventeenth century, Milton continued the tradition when in *Paradise Lost* he writes of the time

... when with fierce winds Orion arm'd
Hath vexed the Red-sea coast, whose waves
O'erthrew Busiris and his Memphian chivalry ...

Orion, under the name of Sahu, was one of the most important sky figures of the ancient Egyptians and was regarded as the soul or incarnation of the god of the afterworld, Osiris. On wall reliefs at the Temple of Dendora (p. 24), he is shown journeying through the heavens in his celestial boat, followed by Sothis (Sirius), who is identified as the soul of Isis and is shown as a kneeling cow with a star between her horns. In some of the oldest writings which have come down to us from ancient Egypt, the *Pyramid Texts* of the late Fifth Dynasty, the king is promised a journey to the realms of Orion:

The Great one has fallen ... His head is taken by Ra, his head is lifted up ... Behold, he has come as Orion ... Behold, Osiris has come as Orion, Lord of Wine in the festival ... he who the sky conceived and the dawn-light bore. O King, the sky conceives you with Orion; the dawn-light bears you with Orion ... by the command of the gods do you live ... with Orion you shall ascend from the eastern region of the sky; with Orion you shall descend into the western region of the sky. ...

△ *Fig 8.3 The Belt and Sword of Orion, showing the nebulosity around Zeta Orionis (the star at the left-hand end of the Belt), with the outline of the Horsehead Nebula just below, and (bottom) the extensive nebulosity around the Orion Nebula, Messier 42, in the Sword of Orion*

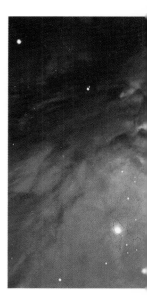

▷ *Fig 8.4 Deep photo of the Orion Nebula, showing a wealth of fine detail*

◁ *Fig 8.5 The Orion Nebula with the four stars of the Trapezium*

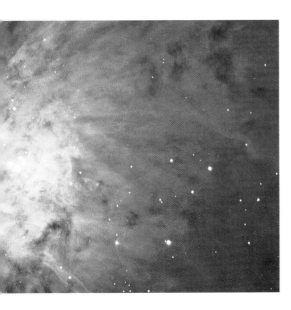

The brightest stars of the constellation are the red supergiant star Betelgeuse – the Armpit of the Giant, which we encountered in chapter 6 – Rigel, the Left Leg of the Giant, and the three stars of Orion's belt, Bellatrix, Mintaka and Alnilam. But the most fascinating object in Orion is the *Orion Nebula*, number 42 in Messier's Catalogue of Nebulous Objects, which may be seen with binoculars as a faint haze spreading out from the quadruple star system Theta Orionis, the central star of Orion's sword. Theta Orionis was first found to be nebulous by Nicholas Pieresc in 1611. The nebula remained relatively unknown till 1656 when Christian Huyghens published the first drawing and description of the object, drawing attention to the remarkable multiplicity of the bright star Theta, in the heart of the luminous cloud. It is in fact four stars, which are known as the *Trapezium*. William Herschel began his first observing journal with an account of observations of the Orion Nebula made with a reflecting telescope of his own construction in 1774, at a time when he was still employed as organist at the Octagon Chapel in Bath. After his brilliant discovery of the planet Uranus in 1781, when he was able to give up his musical duties and study astronomy full-time with the aid of a royal pension, he

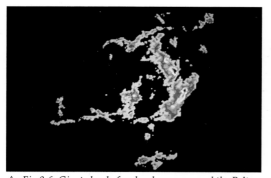

△ *Fig 8.6 Giant cloud of molecular gas around the Belt and Sword of Orion as traced with the molecule carbon monoxide. This cloud weighs 100,000 times as much as the sun*

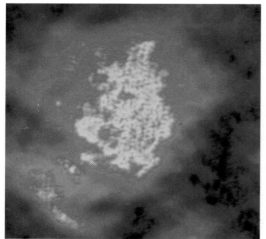

△ *Fig 8.7 Radio picture of the Orion Nebula. Only the hot, ionized gas surrounding the stars of the Trapezium is seen*

△ *Fig 8.8 Near infrared view of the Orion Nebula, showing the locations of the bright stars. The dust is just a haze at these wavelengths*

embarked on a lifelong study of nebulae, thereby laying the foundations of much of modern astronomy. He was to describe the Orion Nebula, prophetically, as 'an unformed fiery mist, the chaotic material of future suns'. His son John Herschel said of the central region of the nebula: 'I know not how to describe it better than by comparing it to a curdling liquid, or to the breaking up of a mackerel sky when the clouds of which it consists begin to assume a cirrus appearance . . .'

In 1864 William Huggins used a prism to break up the light from the nebula into its constituent wavelengths and was able to show that it consisted of a hot, rarefied gas. However it was only with modern studies at infrared and microwave wavelengths that the true nature of the Orion Nebula has become clear. Infrared studies during the 1960s showed that while the nebula is a bright infrared source, other even brighter sources are found nearby with no visible counterparts. In 1968, Eric Becklin and Gerry Neugebauer of the California Institute of Technology found a bright compact infrared source in the Orion Nebula, slightly displaced from the Trapezium. It is now known as the Becklin-Neugebauer object and is believed to be a newly formed, massive star, still deeply embedded in the cloud of gas and dust out of which it has formed. In the same year, Douglas Kleinmann and Frank Low of the University of Arizona discovered an infrared nebula, the Kleinmann-Low nebula, which appears to be a group of massive stars in the act of formation. Then in 1970, Arno Penzias and Bob Wilson, more celebrated for their discovery of the Cosmic Microwave Background (see chapter 21), detected microwave emission at a characteristic wavelength of 2.6 millimetres from the molecule carbon monoxide in the Orion Nebula region. Subsequent studies have shown that the Nebula is embedded in a vast cloud of gas and dust extending over most of Orion's sword. The gas is extremely cold and is mostly in molecular rather than atomic form. The total mass of the cloud is about one hundred thousand times that of the sun. The culmination of these modern studies was the map of Orion made with the IRAS satellite at far infrared wavelengths, which shows both the vast cloud of warm dust and the embedded sources, believed to be newly forming stars. The four stars of the Theta Orionis 'Trapezium' are a small group of luminous stars which formed quite recently out of the cloud, have burned their way out of it, and are now responsible for the illumination of the Orion Nebula. The other newly forming stars have only just switched on their thermonuclear powerhouse and are still heavily shrouded in dust and gas, so that their visible light cannot escape. Instead it is absorbed by the dust and re-emitted at infrared wavelengths. Thus in the infrared we see to the heart of the dense clouds where new stars are forming. IRAS has found many thousands of examples of newly forming stars.

The nebula at the centre of the Sword of Orion opens up an entirely new vision of the stellar universe. Instead of the static distribution of stars in their constellations, we see the cycle of clouds of gas and dust forming, condensing into new stars, and the stars evolving and throwing off shells of gas and dust for the cycle to begin again. As the lifetimes of stars are very long, generally hundreds to thousands of millions of years, the cycle is a slow one in human terms. Even the relatively rapid process of condensation of new stars out of a cloud of gas and dust, which we are witnessing in Orion, takes at least a hundred thousand years.

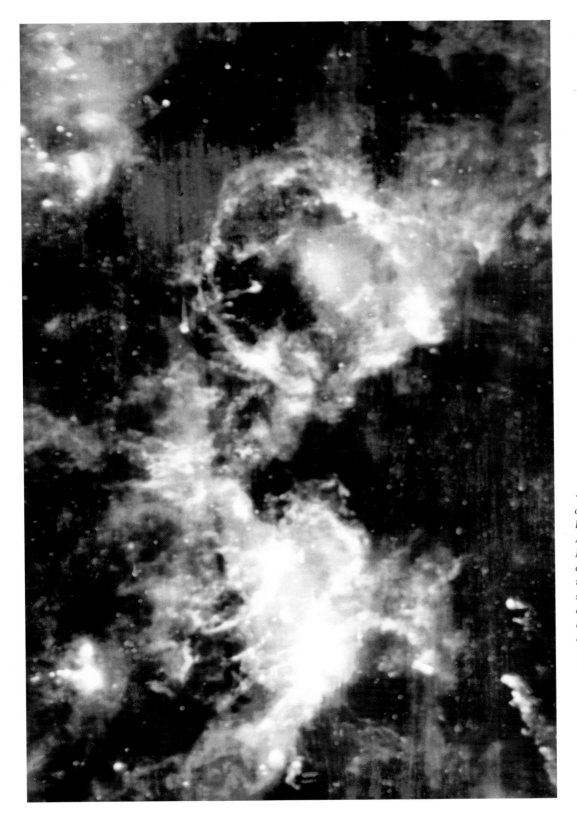

◁ Fig 8.9 Infrared view
of Orion, as seen by the
IRAS Infrared
Astronomical Satellite.
Most of this radiation
comes from dust particles
which have absorbed
starlight. The colour
coding is blue for hotter
dust or starlight, red for
cooler dust. The Orion
Nebula is in the centre of
the bright yellow patch
towards the lower half of
the picture. Notice the
huge ring of dust
surrounding the star
Lambda Orionis, in the
upper part of the picture
and Betelgeuse, the bright
white spot just outside this
ring to the left

△ *Fig 8.10 The dark dust cloud Barnard 84, from a catalogue of such clouds by Edward Barnard*

The existence of material between the stars was more or less unsuspected before 1900. In the early years of the century, Edward Barnard began to study the dark patches that appear throughout the Milky Way and concluded that they were clouds of dust particles absorbing the light from the stars behind, rather than voids in the stellar distribution. One of the most spectacular of these dark patches may be seen with the naked eye in the Milky Way in the southern constellation of Crux, the Cross. Noted in the fifteenth century by the explorers Amerigo Vespucci and Ferdinand Magellan, it is now known as the Coalsack. Microwave studies have shown it to be a vast cloud of dust and molecular gas.

△ *Fig 8.11 The dark dust cloud Barnard 86. A cluster of young blue stars can be seen to the left*

◁ *Fig 8.12 The strange dark outline of the Horsehead Nebula, a dust cloud which has blotted out the light from behind it. The bright white patch to the lower left is a cloud of hot, ionized gas, similar to the Orion Nebula, called NGC 2023*

▷ *Fig 8.13 Infrared view of the Barnard 3 dust cloud, with newly forming 'protostar' (arrowed). The colour coding of this IRAS satellite image is that white is highest intensity, blue is lowest*

Begirt with many a blazing star,
Stood the great giant Algebar,
Orion, hunter of the beast!
His sword hung gleaming by his side,
And, on his arm, the lion's hide
Scattered across the midnight air
The golden radiance of its hair.

LONGFELLOW *'The Occultation of Orion'*

Many a night from yonder ivied casement, ere I went to rest,
Did I look on great Orion, sloping slowly to the west.

Many a night I saw the Pleiads, rising through the mellow shade,
Glitter like a swarm of fireflies in a silver braid.

Here about the beach I wander'd, nourishing a youth sublime
With the fairy tales of science, and the long results of Time . . .

TENNYSON *Locksley Hall*

▷ *Fig 8.14 A visible light picture of the dust clouds surrounding the bright stars Antares (Alpha Scorpii, bottom left), Sigma Scorpii (right) and Rho Ophiuchus (top). Also to be seen in the picture is the globular star cluster Messier 4 (lower right), which lies much further away from us than the other stars. The predominant colours of the three dust clouds (orange, red, blue) are the colours of the stars themselves, seen reflected in the dust clouds, but the effect dust has of changing the colour of light ('reddening') is also apparent*

▽ *Fig 8.15 (top) Infrared view of a larger area of the sky centred on the constellation of Ophiuchus, made with the IRAS satellite. The Milky Way is the bright strip to the lower left. The bright patch in the centre of the picture is the vast dust cloud surrounding Rho Ophiuchus*

▽ *Fig 8.16 (below) Close-up of Rho Ophiuchus in the infrared, showing a newly forming 'protostar'*

◁ *Fig 8.17 Other clouds of hot, ionized gas like the Orion Nebula:*
a The Rosette Nebula
b The Trifid Nebula
c NGC 2264

a

b

c

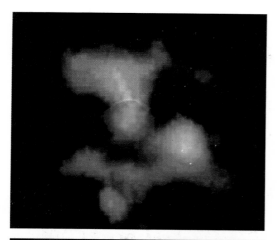

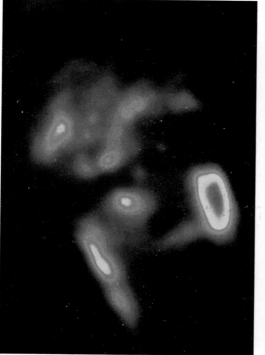

In recent years, the sequence of events when a star forms has begun to be unravelled. A region of above average density in a cloud of gas and dust starts to collapse together under the action of gravity. The collapsing material spins faster, like an ice-skater drawing in her arms, and collapses to form a disc with a central condensation (which will form a star and planetary system) or a disc with several condensations (which will form a multiple star system). Material continues to rain down on the disc and flows in towards the protostar. For reasons not yet understood, the protostar now starts to drive out a high-speed wind, which is channelled by the disc into two conical jets, one from each pole. These *bi-polar outflows* are seen associated with almost every region of star formation. A good example is associated with the very young star T Tauri, which is believed to be a newly formed star of mass similar to the sun. An irregularly variable star associated with a peculiar variable nebula, it was discovered in 1852 by J. R. Hind. Hundreds of such stars are found in the Orion cloud and in the similar cloud in the constellation of Taurus. In such dusty and unusual

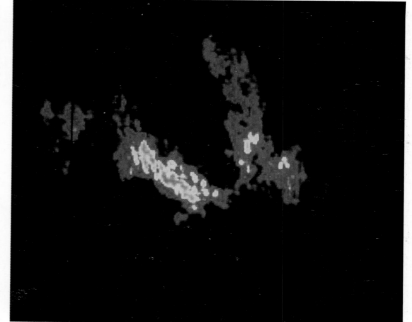

△ *Fig 8.18 The bi-polar nebula Sharpless 106*
(above) Infrared picture, showing the distribution of warm dust (orange) and the central star (blue).
(below) Radio picture. The very faint source at the centre of the two lobes of emission arises directly from the stellar envelope of the star which is illuminating the whole nebula

△ *Fig 8.19 Radio picture of the nebula Messier 17, a region of ionized gas heated by young stars on the edge of a giant cloud of dust and molecular gas similar to the Orion Nebula*

stars we see the last stages of the formation process of a solar system.

The Orion Nebula itself represents a later stage in the evolution of stars much more massive than the sun. Such stars are very hot, many tens of thousands of degrees centigrade, and radiate predominantly at ultraviolet wavelengths. Remember that for radiation from gas, the hotter the material, the higher the frequency at which it radiates. The ultraviolet radiation breaks up the atoms and molecules of hydrogen which make up the bulk of the surrounding gas cloud into their constituent protons and electrons. This process is called *ionization* (p. 54). Hot ionized gas radiates at visible, infrared and radio wavelengths and so wherever hot massive stars form, we see, after the star has had time to burn its way out of its coccoon of dust and gas, bright nebulae which shine at all these wavelengths. Astronomers refer to them as 'HII regions'. Many of these nebulae are rather familiar sights in astronomical books, because they are so colourful. But as we have seen, nebulae like that in Orion are only the tip of a much more interesting iceberg.

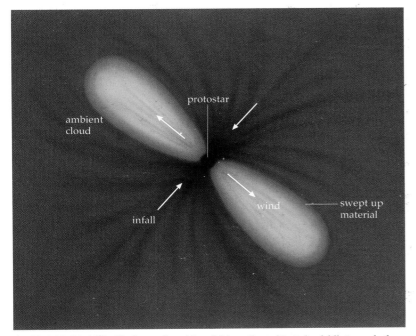

△ *Fig 8.20 Model for bipolar outflow. Material from the ambient cloud falls towards the forming star, or 'protostar', and forms a doughnut-shaped disc around it. This funnels a wind from the star into two oppositely-directed lobes, which create two cavities, surrounded by swept up material*

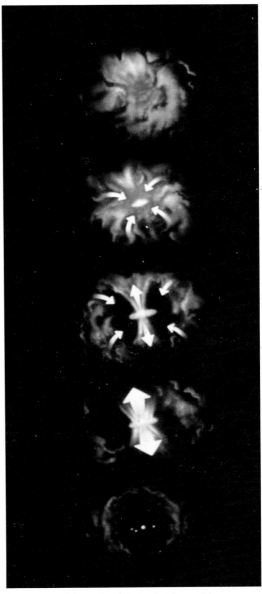

△ *Fig 8.21 Schematic sequence of pictures of the formation of a star (bottom) from a cloud of gas (top)*

C H A P T E R   9

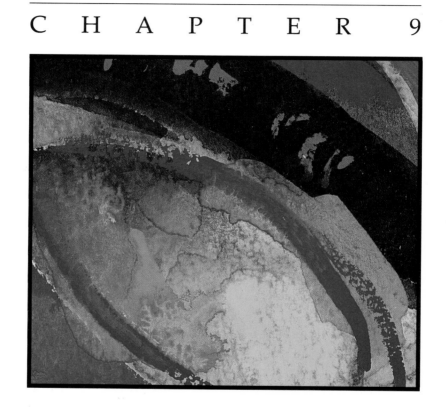

I do not want the constellations any nearer,
I know they are very well where they are,
I know they suffice for those who belong to them.

WHITMAN *'Song of the Open Road'*

DELTA CEPHEI

YARDSTICK FOR THE UNIVERSE

Delta Cephei is a variable star, discovered by the talented young deaf-mute English astronomer John Goodricke in 1784. Goodricke made several important discoveries before his death from pneumonia, almost certainly brought on by long nights at the telescope, at the age of only twenty-two. Delta Cephei varies its brightness with great regularity by a factor of 1.9 every 5.366 days, the rise to maximum taking about one and a half days and the fall to minimum about four days. Its variations are thus faster, more regular, and less dramatic, than those of Mira (chapter 6). Delta Cephei is a yellow supergiant star, at maximum light three thousand times more luminous than the sun. Over five hundred examples of this type of star have been found, and in honour of Delta Cephei they are known as

◁ Fig 9.1 The W of Cassiopeia (upper left) and the kite-shaped constellation of Cepheus (upper right). Delta Cephei is the star immediately to the left of the bottom point of the kite.

Cepheids. The brightest Cepheid in the sky is Polaris, the Pole Star, with a period of variation of just under four days, but its variability is so slight that you are unlikely to notice it with the naked eye. The precisely measurable periods of Cepheids, which range from one day to two hundred days, turn out to be the key to the measurement of distance in the universe. The periodicity of Cepheids is due to regular pulsations of the star's surface.

The colour of Cepheids range from white through yellow to orange and there is a strong relationship between the colour of the stars (which is related to the temperature of the star's surface) and how luminous they are. The more luminous the star, the more yellow or orange it looks. If all stars are plotted in a diagram of luminosity against colour, the Cepheids lie on a strip across the diagram, which is known as the *instability strip*. The instability strip denotes the location of stars which are easily pushed into a state of natural resonance. Bells ring when struck, stretched strings vibrate, and stars pulsate.

In 1912, Henrietta Swan Leavitt unlocked the door to measuring distance in the universe when she found that Cepheids have a very precise relation between their luminosity and their period of variation. Henrietta Leavitt was one of a number of remarkable women working at the Harvard Observatory at the turn of the century as research assistants or 'computers'. The women had to do calculations with hand-operated calculators to analyse the wealth of astronomical observations made at the observatory. In the days before electronic computers and calculators, all such calculations and analyses had to be done by hand.

Amongst the twenty or so Harvard Observatory research assistants and 'computers' of the 1890s, there were at least three who made major contributions to astronomy. Annie Jump Cannon found that the spectra of stars could be arranged in a sequence, which we now know to be one of surface temperature. And Antonia Maury noticed that the widths of the spectral lines in stars of the same types varied from star to star, the first step along the road to the recognition of the distinction between giant and dwarf stars. But the discovery of Henrietta Leavitt was unquestionably the greatest of the three.

△ *Fig 9.2 Henrietta Leavitt*

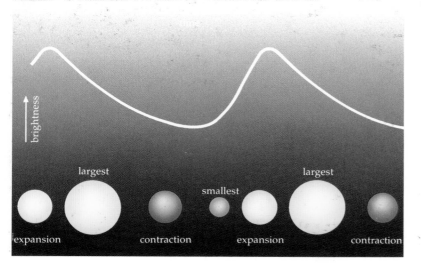

◁ *Fig 9.3 The brightness variations of Delta Cephei with time. This is characteristic of the class of Cepheid variable stars. Over a period from several days to several weeks, the brightness rapidly increases and then gradually fades as the surface of the star pulsates in and out (the changes have been exaggerated in this figure)*

She was studying variable stars in the Magellanic Clouds, large nebulous patches of emission visible to the naked eye in the southern hemisphere. In a large telescope they can be seen to consist of millions of stars and many faint emission nebulae. They will be the subject of chapter 15 of this book. At the turn of the century, the Magellanic Clouds were believed to be part of the Milky Way system. For Henrietta Leavitt's purposes she only needed to know that the stars in the Magellanic Clouds were all at the same distance from the sun. Then differences in the apparent brightness of the stars would correspond to differences in the luminosities of the stars.

She found that there was a rather tight relationship between the luminosity of the variable stars and their period of variation. The longer the period of variation, the more luminous the star. This is known as the *period-luminosity relation* and was announced by Henrietta Leavitt in 1912. Shortly afterwards the Danish astronomer Ejnar Hertzsprung recognized that the variable stars were in fact Cepheids. The significance of this is that if the true distance of a Cepheid, say Delta Cephei, can be established, then we have a beacon with which we can measure distances. We have only to measure the period of variation of the Cepheid and then we can work out how far away the star must be to give the brightness we observe, using the fact that for every doubling of the distance the brightness goes down by a factor of four (*the inverse-square law of radiation*).

There have been some complications along the road to establishing Cepheids as a measuring beacon since Henrietta Leavitt's 1912 discovery. Firstly, it turns out that there are in fact two types of Cepheid variable stars. Type I are the more luminous, longer period Cepheids studied by Henrietta Leavitt. Type II are shorter period variables of lower luminosity and less dramatic variability, like Polaris. However there is a range of periods over which both types are found and here the Type Is are rather more luminous than the Type IIs, for the same period. Confusion between the two types was responsible for an error of a factor of two in the cosmological distance scale, which was only cleared up in 1952 by the American astronomer Walter Baade.

...who sets him
in a constellation and puts the measuring-stick
of distance in his hands?

RILKE *Duino Elegies*

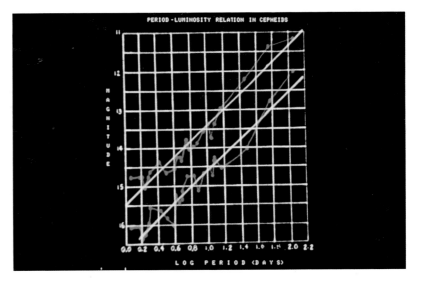

▷ *Fig 9.4 The period-luminosity relation discovered by Henrietta Leavitt for Cepheid variable stars in the Magellanic Clouds*

Another complication is that the light from stars suffers absorption by interstellar dust within our own Milky Way Galaxy and within other galaxies. This was one of the factors that led to a fierce disagreement in the 1920s between Harlow Shapley and Heber Curtis about the size of the Milky Way system (see chapter 14). The problem of working out how much dimming by interstellar dust there has been has inspired a group of astronomers, led by Barry Madore of the University of Toronto, to shift their studies of Cepheids from visible to infrared wavelengths, where the dimming is very much reduced.

We are also left with the problem of determining distances to at least one Cepheid before we can use Henrietta Leavitt's period-luminosity relation. The distances to a few thousand nearby stars can be determined by the method of *parallax*, which we met in chapter 2. As the earth travels round the sun, the direction to a star changes and if this small change can be measured then the distance to the star can be calculated. This is essentially the surveyor's method of triangulation. This does not yet take us out to even the nearest Type I Cepheid, though this important measurement may soon be carried out from space, for example by the Hubble Space Telescope or, if it is relaunched, the ill-fated European Hipparcos mission. At the moment, to measure distances to Cepheids we have to find examples in star clusters and then compare the clusters with a nearby cluster like the Hyades cluster (see chapter 13), for which the distance can be determined by geometric methods. I will return to this question of how astronomers measure distance, which is one that I personally find fascinating, later.

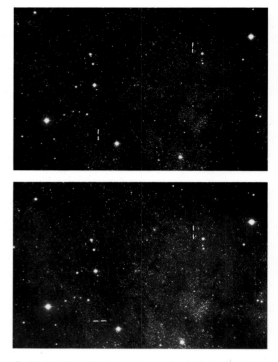

△ *Fig 9.5 Variable stars in the nearby Andromeda galaxy, Messier 31 (see chapter 16)*

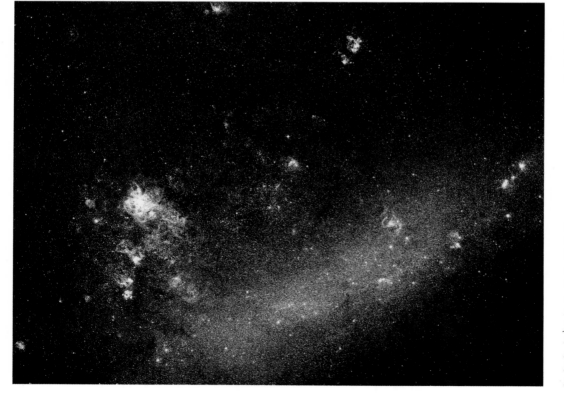

◁ *Fig 9.6 The Large Magellanic Cloud (see chapter 16), in which Henrietta Leavitt was studying variable stars at the turn of the century*

▷ *Fig 9.7 Distribution of stars mentioned in this book in a plot of visual luminosity versus surface colour. The shaded region shows where pulsating stars like Cepheids are found – the 'instability strip'. The solid curve shows where stars still burning hydrogen are located*

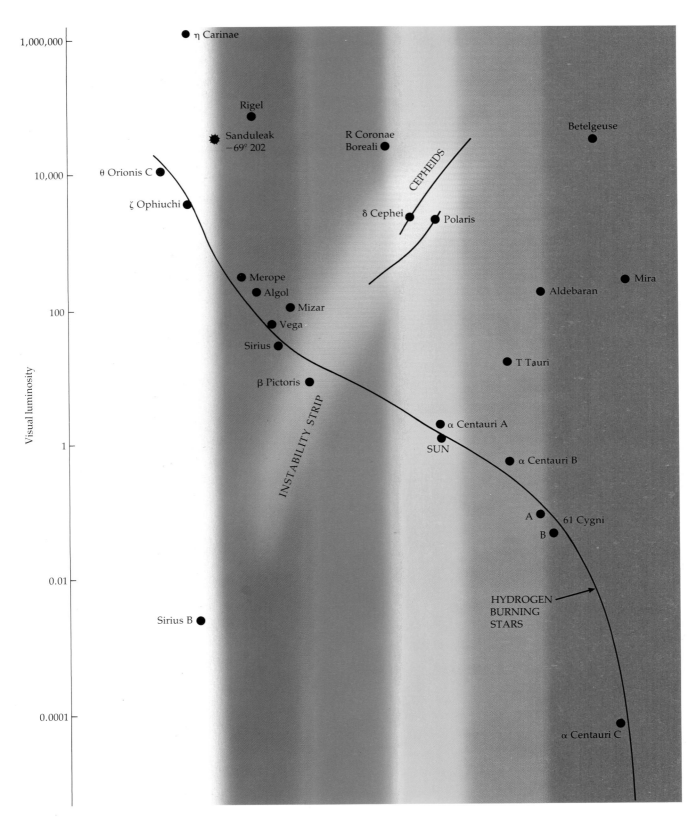

CHAPTER 10

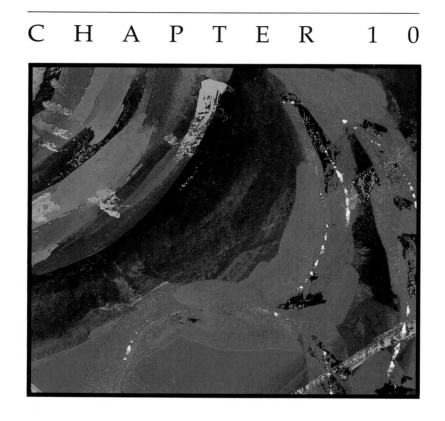

Two stars coming together – a great meeting

OSIP MANDELSTAM *'The Slate Ode'*

ALGOL

THE DEMON STAR

Beta Persei, the demon star, has its name from the Arabic Al Ra's al Ghul, the Demon's Head, the head of Medusa which is held aloft by Perseus. The Perseus story is one of the richest of the Greek myths. Conceived when his mother Danae was visited by Zeus in the form of a shower of gold, mother and son were imprisoned in a wooden chest and cast into the sea in order to evade a prophecy that Perseus would kill his grandfather, King Acrisius of Argos. Luckily the chest floated to the island of Seriphus, where it was found by a fisherman, Dictys, brother of Polydectes, the king of the island. Here Perseus grew to manhood.

It was at the request of Polydectes that Perseus undertook his most famous exploit, the slaying of the loathsome Medusa, one of the three Gorgons whose hair consisted of hissing snakes and whose glance turned people to stone. Equipped with winged sandals, sword and a helmet of invisibility supplied by Athena and Hermes, Perseus found his way to the home of the Gorgons on the farthest shores of Oceanus, near the isles of the Hesperides, and slew the monster while looking at her reflection in his polished shield. From the Gorgon's blood sprang the winged horse, Pegasus. While he was returning from this adventure, Perseus found the princess Andromeda chained to a rock on the Ethiopian coast as a sacrifice to the sea-monster Cetus, sent by the sea-god to punish the kingdom for the boastful vanity of Queen Cassiopeia. Perseus rescued Andromeda and killed the sea-monster. Returning to the court of Polydectes, he turned the scheming king and his noblemen to stone by showing them the Gorgon's head. He also used the latter to relieve the titan Atlas of his wearisome task of holding up the Heavens, transforming him to the Atlas mountains of Morocco. Finally, at funeral games in Thessaly, Perseus fulfilled the prophecy made at his birth when he accidentally killed Acrisius with a discus.

Many of the characters of the Perseus legend appear in the sky as constellations: Perseus himself, Pegasus, Andromeda, Cetus the sea-monster, Queen Cassiopeia and her husband King Cepheus. The vast constellation of Perseus shows him standing holding the Medusa's head aloft with one hand and his sword with the other, in much the same pose as he is portrayed in the great statue by Benvenuto Cellini in the Piazza della Signoria, in Florence.

Algol's name, and its reputation amongst medieval astrologers as the most dangerous and unfortunate star in the heavens, suggest that its strange light changes may have been known to the Arab astronomers of the so-called Dark Ages. The first definite statement on the subject was by the Italian astronomer Geminiano Montanari of Bologna, about 1667. The young John Goodricke first showed the regularity of its period in 1782. He also gave the correct explanation of its periodic dimming as due to the partial eclipse of the star by a dark companion orbiting around it. Every two days, twenty hours, forty-eight minutes and fifty-six seconds, Algol rapidly dims to forty-four per cent of its normal brightness. The eclipse lasts about ten hours. The orbital motion of the primary star was observed in 1889 by H. C. Vogel at Potsdam through the changes in wavelength of its spectral lines due to the Doppler shift (p. 77). The relatively dark companion has never been seen visually because it is lost in the rays of the brighter primary star. It was finally detected spectroscopically in 1978 at the McDonald Observatory in Texas. When the light from a binary system like Algol is passed through a prism spectrometer, spreading the different wavelengths out into a spectrum crossed by bright emission and dark absorption lines, two sets of lines are seen, one from each star. As one star approaches us and the other recedes, the wavelengths of the spectral lines from the two stars are shifted in opposite directions by a small amount, due to the Doppler shift. From the combined results of studies of the light variations, velocity shifts and spectra of the two stars, a detailed picture of the star has emerged.

△ Fig 10.1 The constellation of Perseus, as depicted by the 17 Algol is the bright star in the left eye of Medusa

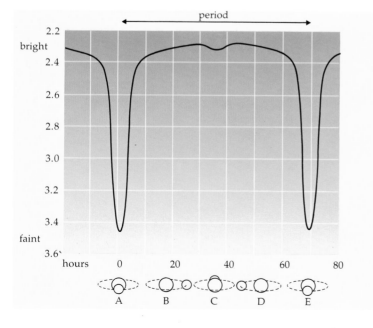

△ Fig 10.2 The binary star system 61 Cygni

◁ Fig 10.3 The light variations of Algol, showing a 69-hour periodicity. The variations can be explained as the alternate eclipsing of a bright supergiant star and a fainter star by each other

tury Swiss astronomer Hevelius.

DOUBLE STARS

Two stars which are close together and orbiting around each other are called a double star or *binary* system. Unless they are being affected by some other nearby star, the two stars will always orbit in the same plane in space, and each star will travel in an ellipse, with the centre of mass of the system as one focus. If the plane of the orbit lies close to our line of sight to the stars, we may see the stars eclipsing each other (an *eclipsing binary* system). If the stars are bright enough and near enough to us to be seen as two separate stars (a *visual* binary system), we may be able to trace out their orbit on the sky over a period of time. Because the stars are usually moving towards us on one part of their orbit and away from us on another part, their light is Doppler shifted. The characteristic bright or dark lines seen across the spectrum (see p. 54) of the stars will change their wavelength first towards the red end of the spectrum (when the star is receding from us) and then towards the blue end (when it is approaching us). When one star is approaching, the other will be receding. A binary system detected by the wavelength changes of its spectral lines is called a *spectroscopic* binary system.

DOPPLER SHIFT

When an ambulance races towards us with its siren wailing, the pitch of the siren is raised to a higher frequency. As it draws level with us we hear the correct note. Then as it hurtles away from us the pitch is lowered. This is the Doppler shift. An approaching source has its frequency raised, or its wavelength shortened (the same thing), a receding source has its frequency lowered, or its wavelength lengthened. The Austrian physicist Christian Doppler demonstrated this effect of a moving source on wavelength or frequency in a spectacular way in 1842, with a brass band playing in a moving railway wagon.

For visible light, longer wavelength means the red end of the spectrum (see p. 11) so the visible spectral lines (see p. 54) are shifted in wavelength towards the red end of the spectrum. We say they are *redshifted*.

Because the amount of the wavelength shift is proportional to the velocity of the source, we can use the Doppler shift to measure the velocities towards us (blueshift) or away from us (redshift) of stars, of gas between the stars, and of whole galaxies. The night sky suddenly throbs with motion.

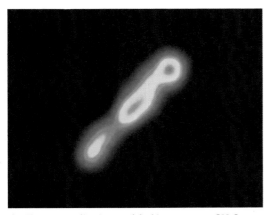

△ *Fig 10.4 Radio picture of the binary system CH Cyg, showing a double-sided jet of emission, in which the material is moving away from the stars at a speed of 1100 kilometres per second.*

The distance of Algol is about a hundred light years. The primary star is a white hydrogen-burning star about a hundred times the solar luminosity and close to 2.6 million miles in diameter. The mass is probably between three and a half and four times the mass of the sun. The 'dark' companion is just a little fainter than the sun, probably of an orange colour, with a diameter of about three million miles and a mass about that of the sun. The separation of the two stars is about 6.5 million miles, so the stars are nearly touching each other. The proximity of the two stars to each other means that they must distort each other's shape very markedly.

Binary stars have fascinated astronomers for centuries. The first really systematic study of binaries was by William and Caroline Herschel at the end of the eighteenth century. They compiled a list of close pairs of stars, the majority of which are true binary systems orbiting around each other, rather than chance superpositions of two stars along the same line of sight. Their catalogue of binary systems, published in 1782, contained 269 binaries. For the next twenty years the Herschels continued to make regular observations of some fifty of these systems and in 1803 William Herschel published the results of their studies. For the binary system Castor (one of the celestial Twins) and for two other systems, the observed relative motions of the stars were good enough to determine the period of their orbits around each other. This was 342 years in the case of Castor. Binary systems where both stars can be seen separately (normally only with the aid of a telescope) are called *visual binaries*. The first example recorded was Mizar, Zeta Ursa Majoris, one of the stars of the Plough or Big Dipper, found to be double through telescopic observations by Riccioli in 1650 (p. 36). The second category, which like Algol require detailed study of their brightness variations, are the *eclipsing binaries*. The plane of the binary system's orbit has to be tilted towards the line of sight for us to see eclipses. The third way that a star can be shown to be a binary system is through spectroscopic observations, as explained above. A binary system found in this way is called a *spectroscopic binary*. Many binary systems fall into two of these categories. Algol is both an eclipsing and a spectroscopic binary. Mizar is both a visual and a spectroscopic binary.

The most exciting type of binary system is one in which the stars are close together, as in Algol. This is because the stars may then interact with each other in dramatic ways. The first indication that this could happen came from Algol itself. Recall that the primary star is a massive supergiant star which is still burning hydrogen: such a star could only be about a hundred million years old. The secondary star, on the other hand, is a solar mass star of large radius, already well on its way towards becoming a red giant. The sun will have this appearance when it is about twice as old as it is now, that is, when it is ten thousand million years old. Yet surely the stars of a binary system must have been born together?

The resolution of this paradox was first proposed by John Crawford in 1955. The secondary star is not what it seems. It looks like an older version of the sun. In reality it was once a much more massive star, probably even more massive than the Algol primary. It completed its hydrogen-burning phase and started to try to become a red supergiant. As it tried to grow in size, it found its shape being distorted by its companion. It became more and more pear-shaped, with the pointed end towards its companion. Then material started to pour through this point, called the Lagrange point after the eighteenth century French mathematician Joseph-Louis Lagrange, onto its companion. The final outcome of this interaction was the system we see today. The star that was originally the primary has dwindled to be only one solar mass. The star that was originally the lower-mass secondary has become so bloated that today it is the primary. In 1955 this was an ingenious but highly speculative explanation for the Algol system. But only two years later, on 4 October 1957, the Soviet Union launched Sputnik I and thus opened up new vistas for astronomy. In particular it became possible to study the universe at X-ray wavelengths, with enormous implications for the study of binary systems.

The birth of X-ray astronomy took place in 1948, when the American Robert Burnight led a team which launched above the earth's atmosphere a photographic film wrapped in silver foil in the nose cone of a captured German V2 rocket. The film became blackened, which showed that the sun emitted X-rays. The next significant development was the launch of an X-ray detector by a team led by Riccardo Giaconni at the American Science and Engineering Corporation in 1962. The purpose of this was to try to detect X-rays from the moon. In this it failed, but succeeded much more dramatically elsewhere. A very bright X-ray source was detected in the constellation of Scorpio. Dubbed *Sco X-1*, it was some time before its nature

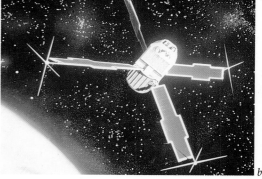

△ *Fig 10.5a Launch of the Einstein X-ray Observatory in 1978*
b The Uhuru X-ray satellite, launched in 1970, which made the first survey of the sky at X-ray wavelengths

▷ *Fig 10.8 Artist's impression of the X-ray binary XB 1820–30, which lies in a globular cluster. A white dwarf orbits a neutron star, which is surrounded by a disc of gas. The Sun (bottom), Earth and Moon (top) are shown on the same scale*

△ *Fig 10.6 Model for X-ray binary system, Cyg X-1. A supergiant star is losing mass to an orbiting black hole surrounded by a disc of accreted gas*

▽ *Fig 10.7 Relative sizes of earth, white dwarf and neutron star*

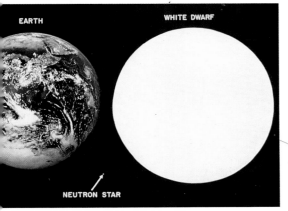

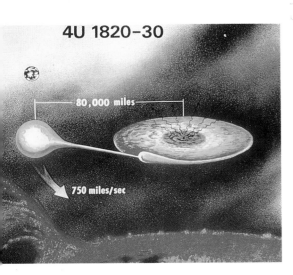

could be identified. Other equally mysterious X-ray sources were found in further rocket experiments, including, in 1965, perhaps the most interesting X-ray source of all, *Cyg X-1*. Then in 1970 a consortium at the Harvard-Smithsonian Center for Astrophysics, again led by Giaconni, launched the Uhuru X-ray satellite from an island off the coast of Kenya. This was a much more sophisticated X-ray detector, which could measure the positions of sources in the sky with some precision. As a result, some of the sources found in the rocket surveys could be identified with stars. In each case it was clear that they were binary systems, with one of the stars being extremely compact, for example a white dwarf or a neutron star. Cyg X-1 was identified with a star from the nineteenth century Henry Draper catalogue, HDE226868. Uhuru also found that the source was variable on a very short time-scale, only one-twentieth of a second. Detailed study of this system shows that the primary is a 20–30 solar mass normal hydrogen-burning star, while the secondary is a dark 10–20 solar mass object. The X-ray emission comes from the vicinity of the secondary and is believed to be due to material pouring off the primary and falling on to the secondary, which must be extremely compact both to account for the very high temperature of the gas for it to be emitting X-rays, and the very short time-scale of variation. We have seen that a star like the sun becomes a white dwarf, and ends its days as a black dwarf. This is not a possibility for a star as massive as the Cyg X-1 secondary, however, as its gravity is too strong to permit a white dwarf structure. Astronomers know of only one dark, compact configuration for a very massive star like this, namely a *black hole*. Such a star has no alternative but to keep on collapsing under gravity until light can no longer escape from it. Material can still fall in, though, and will be accelerated to very high velocities as it does so. Any deviation from a perfect straight line infall will lead to heating of the gas to very high temperatures and hence to the emission of X-rays.

Black holes are undoubtedly the most interesting consequence of Einstein's *General Theory of Relativity*, which he published in 1916. The theory is intended to describe situations where gravity is very strong. In most normal situations, in the solar system for example, Newton's law of gravitation is an accurate description and any differences between Newton's theory of gravitation and Einstein's are very small. Even there though, some subtle differences have been detected. According to Einstein, the path of light, instead of being a straight line, is curved in the presence of matter. This bending of light has been observed by studying the directions of stars at visible wavelengths, and of radio sources at radio wavelengths, as the sun passes near them. A related effect is the slowing down of radar signals reflected off the planets. The orbits of the planets are also slightly modified, an effect which is most marked for Mercury. Although the orbits of the planets are, apart from Pluto, nearly circular, they are in fact, as discovered by Kepler at the beginning of the seventeenth century, ellipses. The major axes of these (almost circular) ellipses slowly rotate or *precess* around the planes of the orbits, mainly due to the effects of the other planets. For Mercury, however, part of the precession remained unexplained from the early nineteenth century onwards and was attributed to an undiscovered planet, Vulcan, lying between Mercury and the sun. This anomaly turned out to be an effect of General Relativity and was triumphantly explained by Einstein in 1916.

The idea of a massive dark object from which light could not escape was first explored by Pierre Simon, Marquis of Laplace, in 1796. He called them 'corps obscurs' – dark bodies. Shortly after the publication of Einstein's General Theory of Relativity, Karl Schwarzschild showed mathematically that an idealized mass concentrated to a point would be surrounded by a 'horizon', from within which light could not escape. The practical significance of this was not clear though. Then in 1939, Robert Oppenheimer and Hartland Snyder showed that a cold and sufficiently massive star would collapse indefinitely. What they realized was that if matter, say the material of a star, is collapsing together spherically, then there eventually comes a moment when the gravity around the star is so strong that light can no longer escape.

A simplified way of thinking about this is to consider what velocity has to be given to a rocket for it to escape from the earth. Launched straight upwards, a rocket has to travel at just over eleven kilometres per second to escape from the earth's gravity. This is called the *escape velocity* from the earth. The escape velocity from the surface of the sun is much larger, 617 kilometres per second. From the surface of a white dwarf, the escape velocity is ten thousand kilometres per second. But if a star has collapsed to a state where the escape velocity from the surface exceeds the velocity of light, then no rocket nor any cry for help can escape from the star. It has become a black hole. It is invisible to sight, though if we could orbit around it at a safe distance, we would still feel its gravity.

In the 1960s the idea of black holes was intensively investigated and finally in the early 1970s Roger Penrose, Steven Hawking and other relativists proved that there was no escape from the final black hole state for a collapsing massive star.

After that detour around black holes, let us return to the question of the nature of Cyg X-1. Since a black hole is invisible, how can we know that a distant X-ray source like Cyg X-1 is one? Firstly there is a process of elimination. We know of only two stable collapsed states for dead stars, white dwarfs and neutron stars, which we will encounter in chapter 12 (they are the relics of supernova explosions). Both of these are limited to objects no more massive than about two times the mass of the sun. Thus if we find a compact, dead, dark object much more massive than this, it probably has to be a black hole. More directly we can probe the region close to the black hole, where the enormous strength of gravity causes very high infall speeds, close to the speed of light. Thus any evidence for particles being accelerated to approximately the speed of light may help to confirm the black hole diagnosis. Finally, a source cannot change its luminosity significantly in a time shorter than it takes for light to cross the source, so ultra-rapid variations can demonstrate that a source is so compact that it can only be a black hole. All three of these diagnostics are present in the case of Cygnus X-1.

Hundreds of examples of X-ray emitting binary stars are now known. In

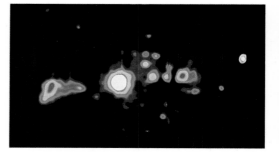

△ *Fig 10.9 SS433 at radio wavelengths*

The clear sky is strewn with horrible dead suns
Dense sediments of crushed atoms:
Nothing emanates from them but desperate heaviness,
No energy, no messages, no particles, no light;
Light itself falls back, broken by its own weight.

PRIMO LEVI *'The Black Stars'*

▷ *Fig 10.11 Lawn-sprinkler model for SS433. Material from the visible companion star falls onto an accretion disc surrounding a dense invisible component, which emits two rapidly moving emission beams. The complex spectrum is a result of the combined orbital motion and the precession of the two beams*

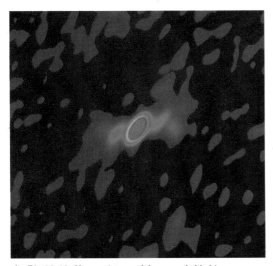

△ *Fig 10.10 X-ray picture of the remarkable binary system SS433*

each case there is a primary star losing mass and a compact secondary star, which could be a white dwarf, a neutron star or black hole. Amongst the stars visible to the naked eye which are X-ray binaries are Beta Lyrae, a star which is eclipsed by an extended disc of material surrounding a dark, massive object which may be a black hole, and Epsilon Aurigae, also eclipsed by a disc of material around an unseen star.

In 1978 a binary system was discovered which for a while caused a furore in astronomical circles. It is known as SS433, because it is number 433 in a catalogue of stars with unusual spectra made by the American astronomers Bruce Stephenson and Nicholas Sanduleak in the 1960s. It was discovered as a result of a search for the progenitor of the radio supernova remnant W50, number 50 in the catalogue of extended radio sources in the Milky Way made by the Dutch radio-astronomer Gart Westerhout. The link between the huge radio source and the unusual star was provided by the X-ray survey carried out by the British Ariel 5 satellite, which found an X-ray source near to a strong radio peak at the centre of W50.

In June 1978 two British astronomers, Paul Murdin and David Clark, working at the Anglo-Australian telescope at Coonabarrabrand in Australia, identified the X-ray source with SS433 and found that the star showed three sets of spectral lines (p. 54). One group was redshifted, a second group was blueshifted and a third group showed no shift at all. The two shifted groups moved backwards and forwards across the spectrum in phase with each other every thirteen days, and implied motions close to the speed of light. The star also showed variability on a time-scale of 164 days. Unfortunately Murdin and Clark were unable to pursue their observations for long enough to unravel what was going on, and it was left for Bruce Margon of the University of California at Los Angeles to solve the mystery. He was able to follow the variations of the two moving sets of spectral lines through several cycles and from these observations emerged a picture of what was happening. SS433 consists of two stars in orbit around each other once every thirteen days, a luminous visible star with a rotating neutron star companion surrounded by a disc of material accreted from the visible primary. The neutron star is the remnant of the supernova explosion which produced the radio source W50 and the accretion disc is the origin of the X-ray emission. Along the axis of the neutron star, two jets of material are being ejected continuously at very high velocity and it is from these beams that the moving sets of spectral lines emanate. Tidal interaction between the two stars causes the neutron star's rotation axis to precess once every 164 days. The system is not that different from other X-ray binaries, apart from the remarkable beams. Why this neutron star should eject two beams instead of producing a pulsar (see chapter 12) like other neutron stars is something of a mystery.

The complexities of binary star interactions are continuing to be unravelled and perhaps new kinds of systems remain to be discovered.

C H A P T E R   1 1

And new fixt starres found in that Circle blue,
The one espide in glittering Cassiopie
The other near to Ophiuchus thigh.
Both bigger than the biggest stars that are,
And yet as farre remov'd from mortall eye
As are the furthest, so those Arts declare
Unto whose reaching sight Heavens mysteries lie bare.

. . . Neither did last the full space of two year.
Wherefore I cannot deem that their first day
Of being, when to us they sent out shining ray.

HENRY MORE *'The Infinity of Worlds'*

NOVA AQUILAE

THE NEW STAR OF 1918

On the night of 8 June 1918 several people independently noticed that a bright new star had appeared in the constellation of Aquila. Among the earliest was the astronomer Edward Barnard, whom we have already met as the discoverer of interstellar dust, and a youth of seventeen who was to become America's leading amateur astronomer and comet-discoverer, Leslie C. Peltier. Within hours the new star, or *nova*, had become as bright as Sirius, the brightest star in the sky. Examination of photographs taken in the same region of the sky a few days earlier showed that the star brightened by a factor of one hundred thousand in six days. This was the brightest nova seen for three hundred years.

It was the discovery of a nova in the constellation of Scorpius in 134 BC which inspired Hipparcos to embark on the massive task of compiling his star catalogue, a work which proved to be immensely valuable for over 1700 years. But even more remarkable were the achievements of the ancient Chinese astronomers. Every night for thousands of years the Chinese Imperial astronomers gathered on the roof of the Imperial Palace. One watched towards the north, one to the south, one to the east, one to the west, and a fifth looked upwards towards the zenith. We know this because a seventeenth century Italian missionary saw them in action. As a result of this extraordinary vigilance, they were familiar with all the changing phenomena of the night sky, phenomena which western civilization, blinded by the mistaken genius of Plato and his pupil Aristotle, did not begin to notice properly till the time of the Renaissance. Variable stars, meteor streams, comets, were all routinely recorded. And, most impressive of all, they recorded the appearance of new stars so precisely that we are still able to use their observations for scientific work today. The oldest surviving record of a new star, or *nova*, is an oracle-bone dating to about 1300 BC. The inscription on the bone says: 'On the 7th day of the month, a chi-ssu day, a great new star appeared in company with Antares.'

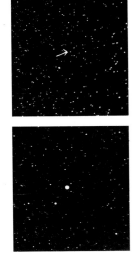

△ *Fig 11.1 Nova Aquilae 1918 before and after outburst*

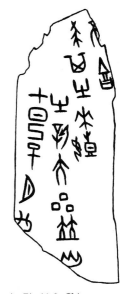

△ *Fig 11.2 Chinese oracle-bone inscription from about 1300 BC describing the appearance of a new star or nova*

Some of these 'guest stars' were the more dramatic phenomenon of supernovae which we will meet in the next chapter. But as centuries can pass with no new supernova visible to the naked eye, the majority of the Chinese guest stars were undoubtedly novae, which tend to occur every ten years or so.

After reaching its maximum brightness on 9 June 1918, Nova Aquilae began to decline, but not as rapidly as its increase. It took a month to decline by a factor of a hundred, and seven years to return to its normal, quiescent state. Seventeen novae were recorded in our Milky Way Galaxy between 1875 and 1975, the most recent occuring in Cygnus and reaching its maximum on 29 August 1975.

In modern times we have come to understand the cause of this spectacular phenomenon. Novae occur in close binary systems in which one of the stars is a white dwarf. Now we saw in the last chapter that when two stars are very close together in a binary system, it is common for material to be stripped off one star and deposited on the other. In the case of novae, hydrogen from the normal companion star falls on to the white dwarf at great speed and becomes so hot that nuclear reactions begin in a sudden burst on the surface of the white dwarf. A rapidly moving shell of gas is then ejected from the white dwarf in all directions. This is the shell of gas we see on optical photographs of novae. Although violent, the nova event does not disrupt the white dwarf star and it returns to its quiescent state.

The nova T Corona Borealis (bright variable stars are denoted by one or two capital letters and the name of the constellation in which they occur) erupted both in 1866 and again in 1946, when it was a moderately bright star at maximum, and several other examples of *recurrent novae* are known. RS Ophiuchi erupted in 1898, 1933, 1958 and again in 1985. The explanation for these is similar to that for 'classical' novae like Nova Aquilae, except that the event is less violent. On an even more frequent and weaker scale is the *dwarf nova*, U Geminorum, which brightens by a factor of a hundred every few months. A normal red star deposits gas on to a disc surrounding a white dwarf companion, which heats up and brightens. As the system is viewed edge-on, the red star and disc eclipse each other, but during outburst no eclipse is seen. Such systems have been the subject of intensive study in the past twenty years because they are relatively easy to observe and a great deal of information can be accumulated about each outburst. As a result, enormous insight into the way close binary stars interact with each other has been gained, which helps us to understand the rarer and more violent nova phenomenon and the even rarer and even more violent supernovae. Because novae outbursts follow a definite pattern of brightening and dimming, they can be used to estimate the distances to galaxies beyond our Milky Way system. Recently Canadian astronomers Sidney van den Bergh and Christopher Pritchet detected novae in the Virgo cluster of galaxies (chapter 20), at a distance of sixty million light years.

The sun will eventually become a white dwarf, when it has completed its evolution as a red giant and thrown off its outer layers as a planetary nebula. It has no companion to wake it up by depositing gas on its surface, however, so it will never appear as a nova to some watcher circling a distant star.

△ *Fig 11.3 Nova Persei 1901 showing filaments thrown out by the explosion*

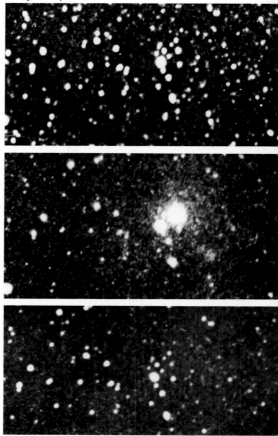

△ *Fig 11.4 Nova Lacertae 1910 before (top), during and after outburst*

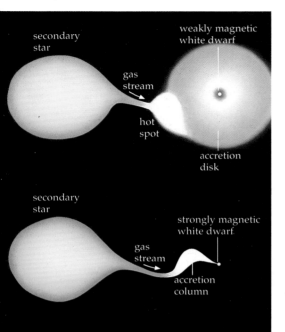

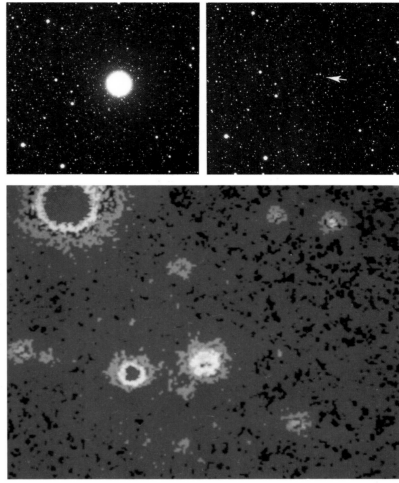

△ Fig 11.5 Two possible models for dwarf novae, also known as cataclysmic variables. In each case a white dwarf is in orbit with a normal secondary star which is trying to grow in size to become a red giant. As the secondary swells, gas overflows from it and streams towards the white dwarf. If the dwarf is only weakly magnetized, the gas forms a hot, thin gaseous disc. A strongly magnetized dwarf will prevent a disc from forming: instead the gas stream is channelled directly on to the surface of the white dwarf

△ Fig 11.7 (top) The most recent bright classical nova, Nova Cygni 1975 during and after the outburst. The star brightened by a factor of forty million

△ Fig 11.8 (below) First detection of the debris from the Nova Cygni outburst, nine years later in 1984. The shell of gas (right-hand image) is seen in the light of the H-alpha hydrogen line

△ Fig 11.6 Model for classical novae. Material fed to an accretion disc from a companion (as in the model for dwarf novae) falls onto a white dwarf, is heated to a high temperature and ignites in a dramatic thermonuclear flash which drives out a shell of gas

◁ Fig 11.9 Radio picture of the 1985 outburst of the recurrent nova, RS Ophiuchi, made with a network of radio-telescopes all over Europe. The outburst appears to have been in the form of twin, oppositely pointing jets rather than a spherical shell. The map shows a region around the erupting star only about five times larger than the orbit of Pluto around the sun

C H A P T E R   1 2

Behold, directly overhead, a certain strange star was
suddenly seen . . . Amazed, and as if astonished and
stupefied, I stood still.

TYCHO BRAHE *1572*

CRAB NEBULA

THE SUPERNOVA OF AD 1054

The Crab Nebula was discovered in 1731 by the English physician and amateur astronomer John Bevis and, independently, by the French comet-watcher Charles Messier, twenty-seven years later. It was his discovery of the Crab Nebula which made Messier realize that he had to do something about the many bright nebulous objects that could be mistaken for new comets. He therefore embarked on his famous *Catalogue of Nebulous Objects* – objects for comet-watchers to avoid but of the greatest interest to all other astronomers (p. 51). The Crab Nebula became number 1 in Messier's *Catalogue*. The nickname the 'Crab' derives from the appearance of the complex, twisted filaments. In this century these filaments have been seen

◁ *Fig 12.1 The Crab Nebula in visible light. This remnant of the supernova of AD 1054 is now 6 light years in diameter and still spreading outwards*

△ Fig 12.2 The Crab Nebula at radio wavelengths, one of the brightest radio sources in the sky

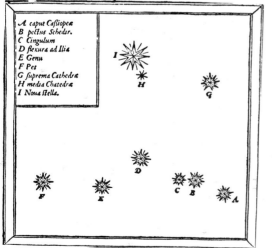

△ Fig 12.3 Navaho Indian petroglyph, which may depict the AD 1054 supernova which gave rise to the Crab Nebula

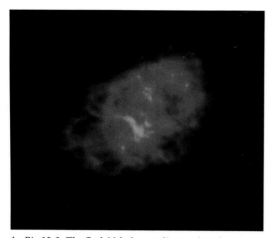

A caput Caſsiopeæ
B pectus Schedir.
C Cingulum
D flexura ad Ilia
E Genu
F Pes
G ſuprema Cathedra
H media Chatedra
I Noua ſtella.

△ Fig 12.4 Tycho Brahe's sketch of the supernova he studied in 1572. The 'nova stella' is labelled I and the stars A–G make up the 'W' of Cassiopeia

to expand outwards, and in 1942 the American astronomer Walter Baade estimated an age for the cloud of 760 years (now revised to 900 years). The chronicles of medieval China contain an account of the original event which caused the nebula:

> In the 1st year of the period Chih-ho, the 5th moon, the day chi-ch'ou, a guest star appeared approximately several inches south-east of Tien-Kuan . . . After more than a year it gradually became invisible (*Annals of the Sung Dynasty*)

The precise date, according to this account was 4 July 1054, and the position, south-east of the star Zeta Tauri, is that of the Crab nebula today.

It is surprising that this dramatic event, the appearance of a bright new star, does not appear to have been noticed in Europe. It is possible, however, that it was seen by the Navaho Indians of Northern Arizona. A carving in the rock wall of a cave depicts a bright star close to a crescent moon. It is a fact that the new star of AD 1054 did appear near the crescent moon. Could the Navaho have recorded the event which gave birth to the Crab nebula? As there is no written record, we shall probably never know. It is only fair to point out that a star and crescent symbol is not rare in history, having been used in Sumeria, ancient Rome, and, of course, Islam.

We saw in chapter 11 that the appearance of a new star or nova in the sky is not that rare an event. Novae like Nova Aquilae tend to recur every few decades, brightening a million times for a few months and then dimming again. In 1934 Walter Baade and Fritz Zwicky realized that certain nova events are, however, quite different. In these 'supernova' events a star brightens as much as a thousand million times in a dramatic convulsion that signals the destruction of the star. The Crab nebula is the relic of just such an event and is the best studied of all supernova remnants.

Several other supernova events have been recorded in human history. The Chinese astronomers recorded several, including those of AD 185,

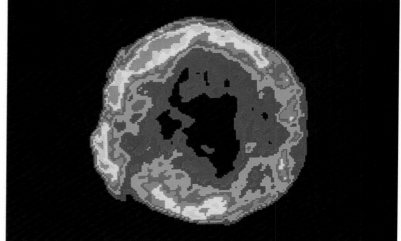

△ Fig 12.5 A colour representation of the remnant of Tycho's supernova, as it appears today at radio wavelengths. The bright ring is the surface layers of the star still expanding outwards

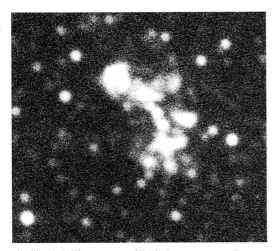

△ *Fig 12.6 The remnant of Kepler's supernova as it appears today at visible wavelengths: only part of the shell is illuminated*

AD 386, AD 393, AD 1006 and AD 1181. That of 1006 was also seen by Swiss monks.

Two of the greatest astronomers of the Renaissance were lucky enough to see supernovae. The great Danish astronomer Tycho Brahe observed a supernova in Cassiopeia on 11 November 1572 (it seems to have been seen a few days earlier by W. Schuler). Tycho made extremely careful observations of this star, noting its gradual change in brightness from its first appearance, when it rivalled Venus, to its disappearance sixteen months later. He showed that it did not change its position relative to the other bright stars in Cassiopeia and deduced that the new star must be further off than the moon and hence in the region of the so-called uncorruptible and unchangeable fixed stars. Only thirty-two years later, on 9 October 1604, another supernova was seen by several observers. Johannes Kepler studied it on 17 October and found it to be similar to Tycho's star.

The remnants of many supernova events are seen throughout the Milky Way. The hot gas expanding outwards makes an especially brilliant display at radio and X-ray wavelengths.

Although supernova events have followed on each other's heels in a matter of decades more than once in recorded history, it is remarkable that no supernova has been reliably seen in the Milky Way since 1604. Everything astronomers have learnt about supernovae tells them that supernovae should occur in our Galaxy about once every fifty years. We have been exceptionally unlucky not to have one visible from our part of the Milky Way for so long. In fact there may be one other recent historical supernova: in 1667 a star was observed by Flamsteed which may have been the event which gave rise to the supernova remnant now known as Casseopeia A.

This was found first as a very bright radio source, but can be seen too in X-rays and at visible wavelengths. Most of the hundreds of supernova remnants known in the Milky Way are due to events many thousands of years old.

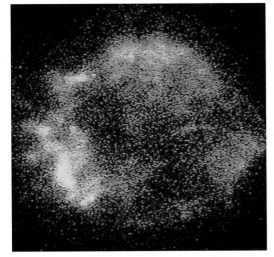

△ *Fig 12.7 Cassiopeia A supernova remnant in X-rays. The emission comes from an expanding shell of gas at a temperature of millions of degrees Centigrade.*

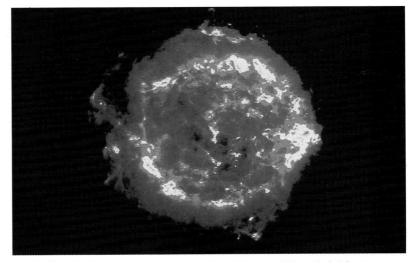

△ *Fig 12.8 Cassiopeia A supernova remnant in radio waves. This is the brightest source of radio waves in the sky. The supernova explosion may have been seen by Flamsteed in 1667*

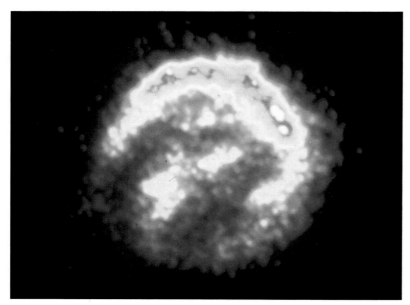

△ *Fig 12.9 X-ray image of the remnant of Kepler's supernova of 1604, a Type I event*

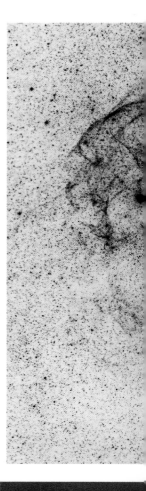

▷ *Fig 12.10 The complex and delicate filaments of a much older supernova remnant, S147, seen in visible light. The filaments cover an area over 40 times that of the moon*

▽ *Fig 12.11 Sketch of the two types of supernovae: (left) Type I: material flows from the large, pear-shaped red giant onto the tiny white dwarf, causing it to undergo a thermonuclear explosion, (right) Type II: the core of a massive star collapses to form a neutron star or black hole, while a shock wave blasts out through the outer parts of the star*

TYPES OF SUPERNOVAE

What is a supernova? The simplest answer is that it is an exploding star, but it turns out that there are at least two completely different types of stellar explosion involved. In either case a star brightens by a huge factor, as much as a billion, in the course of a week, before starting to fade more gradually, taking a year or two to disappear completely.

ORIGINS

The first type of supernova, called Type I by astronomers, originates in a double-star system in which one of the stars has become a white dwarf (see chapter 3). In the course of its evolution the white dwarf's companion star starts to grow in size, but after a while it cannot grow further because of the proximity of the white dwarf. When this point is reached, material starts to flow from the companion star onto the white dwarf. Now white dwarfs are not allowed to be more massive than the 'Chandrasekhar limit', discovered by Subrahmanyan Chandresekhar of the University of Chicago. If a white dwarf is more than about 1.4 times the mass of the sun, a law of quantum mechanics about how tightly matter can be packed together would be violated at the centre of the white dwarf. If the material from the companion takes the white dwarf over this critical limit, the white dwarf undergoes a catastrophic explosion, resulting in the supernova event.

The second type of supernova, known as Type II, has a quite different origin. These supernova events are the death throes of stars much more massive than the sun, say from 8 to 60 times as massive. The late stages of evolution of such a star are very complex, with the centre of the star

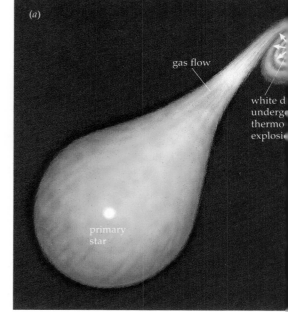

(a)

gas flow

white d
underg
thermo
explosi

primary
star

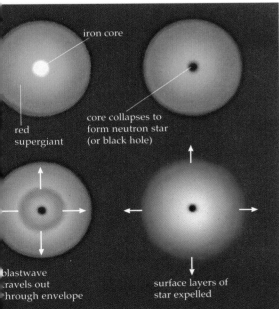

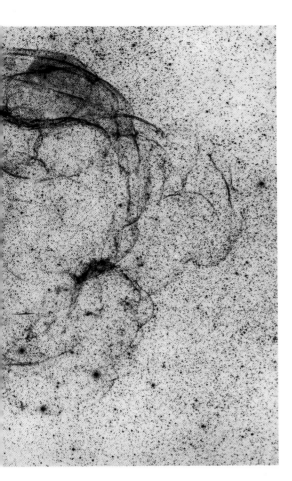

iron core

core collapses to
form neutron star
(or black hole)

red
supergiant

blastwave
travels out
through envelope

surface layers of
star expelled

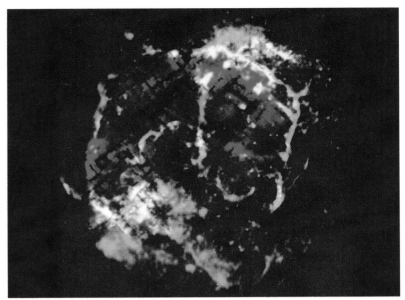

△ *Fig 12.12 An X-ray and optical composite of the Vela X supernova remnant. A Type II event*

becoming progressively hotter, and progressively heavier elements being synthesized in the thermonuclear furnace – helium, carbon, nitrogen, oxygen, silicon, iron. But when the centre of the star is made of iron an energy crisis occurs, since the thermonuclear fusion of iron *absorbs* energy instead of releasing it. The star starts to collapse and the central regions just keep on collapsing together. If this hot iron core is not too massive (not too much heavier than the sun, say), the collapse can eventually be halted when the density becomes so high that all the atomic nuclei are crushed together, forming a *neutron star*. If the core is too massive for the pressure of the tightly packed neutrons to support it against gravity, the core continues to collapse until gravity becomes so strong at the centre that a *black hole* is formed (see chapter 10). The shock of the central regions crashing together reverberates through the more tenuous outer layers, driving them outwards at enormous speed (more than 10,000 km per second) and igniting a whole series of thermonuclear reactions in which the radioactive elements are made. In both types of supernova the slow and regular decline of their visible light output is due to the radioactive decay of cobalt-56 to iron.

There are subtle differences in the appearance of the two types of supernova which allow astronomers to distinguish between them. Type II supernovae show the characteristic emission lines of hydrogen from the exploding star's outer layers in their spectra. They leave a remnant of the stellar core behind, usually a neutron star, and are only found in regions where new massive stars are forming; for example in the arms of spiral galaxies. Type I supernovae show no hydrogen lines and are found both in spiral and elliptical galaxies. They are usually at least twice as luminous, at maximum light, as Type II supernovae. The Crab supernova of AD 1054 was of Type II, the death of a massive star. Those of Tycho and Kepler were of Type I, the explosion of a white dwarf in a binary system.

DISCOVERY OF A PULSAR

In 1967 a discovery was made which suddenly revealed to astronomers the final stages of Type II supernovae events like that of AD 1054. To study the twinkling of radio sources due to the 'solar wind', ionized gas which is driven steadily off the surface of the sun, Anthony Hewish and colleagues at the Mullard Radio Astronomy Observatory, Cambridge, had set up a massive array of 2048 radio aerials covering an area of 4 acres. The task of studying the output from this unusual radio-telescope was given to a young post-graduate student, Jocelyn Bell. In July 1967 she noticed some curious recurring signals on the recording chart when the telescope pointed at a certain direction in the sky, 'scruff' as she called it. She consulted Hewish and they decided to install high-speed data recorders. They found that the mysterious source emitted sharp pulses at very precise intervals of just over one second. The source seemed like a celestial time signal. For a while Hewish and Bell wondered if this was a signal from 'little green men', other intelligent life. What an exciting and terrifying prospect that must have been to them. They soon had to abandon this fascinating idea, though. The source was called a pulsating radio source or *pulsar* for short. Other pulsars were quickly discovered, among them one in the middle of the Crab nebula. This turned out to be one of the most rapidly pulsating of all the pulsars discovered so far, with a period of only one-thirtieth of a second. This very rapid period of pulsation gave the crucial clue to the nature of the pulsating source. Such a regular pulse could only come from a rotating object and the only object which can rotate as rapidly as thirty times a second without breaking apart is a neutron star.

It is a remarkable tribute to human ingenuity that the properties of neutron stars had been worked out decades before they were actually discovered. One of the people who played a key role in this was Robert Oppenheimer, the man who led the Manhattan project to build the atomic bomb. The matter in a neutron star is so compressed that a thimbleful

△ *Fig 12.14 The array of radio aerials with which Anthony*

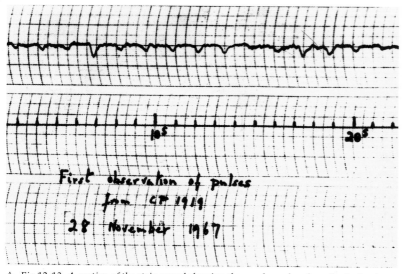

△ *Fig 12.13 A portion of the strip record showing the regular pulses from the first pulsar discovered*

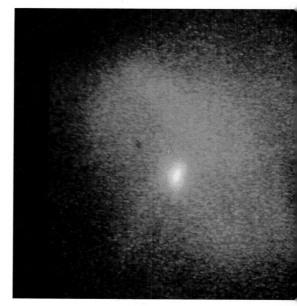

would weigh a billion tons. A neutron star with the same mass as the sun occupies so small a volume that it would not even shade a city like London or New York.

Although pulsars were first detected at radio wavelengths, the same pulses have been seen at other wavelengths in some objects. The Crab pulsar was the first to be detected at visible wavelengths and also the first to be seen pulsating in X-rays. Although the period of pulsation remains amazingly steady, a very small slowing in the period can be measured over a long enough period of observation. The slowing is because the rotation of the neutron star is slowing down as the energy of rotation is used to drive the pulses. The Crab pulsar, only nine hundred years old, is in fact the youngest of all the pulsars we have found in our Galaxy. Other pulsars are over a million years old. Perhaps some of the supernovae which gave rise to them were noticed by our earliest ancestors.

Hewish and Jocelyn Bell discovered the first pulsar

△ *Fig 12.15 Crab nebula in optical light, with position of pulsar indicated by arrow*

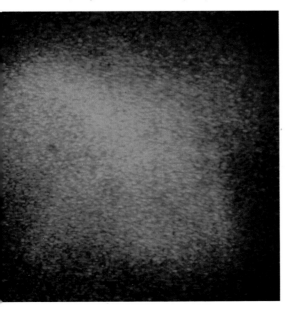

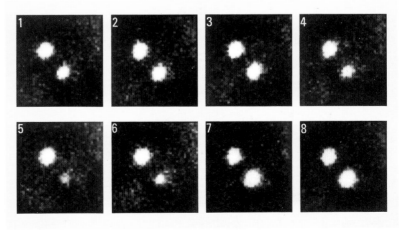

△ *Fig 12.16 A rapid series of exposures in visible light of the Crab pulsar. The pulsar is the object to the lower right and can be seen to be varying in brightness*

◁ *Fig 12.17 The Crab pulsar on and off, as seen in X-rays*

Why do pulsars pulse? Although this is not completely understood, it is believed that pulsars must contain an extremely strong magnetic field, a million million times stronger than that of the earth, and that this controls the shape of the pulse. Two ideas are illustrated in Fig. 12.18, both of which are variants of the 'light-house' model. This term is based on the fact that a ship at sea sees the beam from a lighthouse as a series of flashes, though in fact the beam really sweeps round the sky in a periodic motion. In one idea the magnetic axis differs slightly from the rotation axis (as is the case for the earth) and a beam is directed along the magnetic axis which sweeps regularly past the field of view of our telescopes. In the second idea the rapidly rotating neutron star and its enormously powerful magnetic field drag a cloud of electrically charged gas around with it and the beam is generated out near the zone where the gas is being dragged round at the speed of light.

If astronomers had to make do with the supernovae in our own Galaxy, they would not have learnt much about them yet. Remember that the last definitely observed supernova event in our Galaxy was that studied by Kepler in 1604. Fortunately, many supernovae are seen in other galaxies and from these events a very detailed picture of a supernova explosion can be built up. The largest available computers are needed to calculate the details of what happens.

△ Fig 12.18 *The lighthouse model for pulsars. The beam of radio waves (and visible light or X-rays) sweeps past us each time the star goes through one revolution. The beam may be either along the magnetic axis, or generated at the surface of a doughnut-shaped cloud of electrically charged gas being dragged round with the star*

△ Fig 12.19 *Type I supernova 1972e in the spiral galaxy NGC5253*

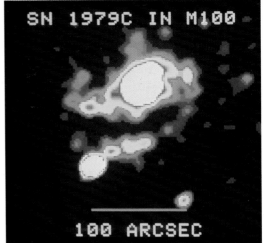

△ Fig 12.21 *Radio picture of Type II supernova 1979c in the galaxy M100. The supernova is situated on the edge of one of the galaxy's spiral arms. Careful study of the expansion of this radio remnant has allowed the distance to this galaxy to be measured accurately*

△ Fig 12.20 *Two views of the peculiar elliptical galaxy NGC5128, which has a pronounced band of dust across it, showing the 1986 supernova*

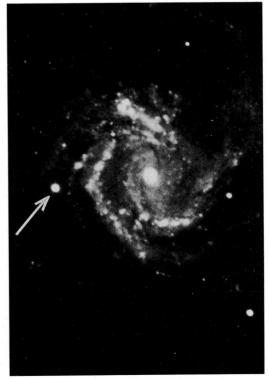

△ Fig 12.22 *Supernova in the spiral galaxy NGC4303 (arrowed)*

△ Fig 12.23 *Supernova 1987a in the Large Magellanic Cloud (the bright star with a cross), the most exciting astronomical event of the century*

THE SUPERNOVA OF 1987

On the night of 23 February 1987 an event occurred which lit up astronomers around the world. A supernova was seen in the Large Magellanic Cloud, a small galaxy which is the nearest neighbour to our own Milky Way galaxy, only 150,000 light years away. Because there are always several astronomers studying this region of the sky any night, several astronomers independently discovered the supernova. Ian Shelton of the University of Toronto, working at the Las Campanas Observatory in Chile, first discovered the supernova at 05.30 GMT on the night of the 24th. For several nights he had been taking long three-hour exposures of the Large Magellanic Cloud to locate faint objects. When developed, that night's photographic plate revealed a brilliant new star, the supernova. At 07.55 GMT one of the observatory's assistants, Oscar Duhalde, was strolling outside and noticed the supernova as he glanced at the Large Magellanic Cloud. One hour later, a seventy-year-old astronomer, Albert Jones, spotted the supernova from New Zealand. He contacted Rob McNaught at the Anglo-Australian Observatory in New South Wales. McNaught had been taking photographs of the Large Magellanic Cloud every night for five months, but for some reason he did not look at the photograph he had taken on the night of 23 February, at 10.35 GMT, until the call came from Jones a day later. When he did so he found that he had the first record of the supernova. By now telegrams and phone-calls were flying round to the observatories of the southern hemisphere, warning them to turn their attention to the supernova, the nearest known since the time of Kepler four hundred years ago.

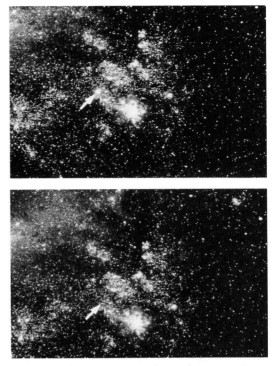

△ Fig 12.24 *The photographs on which supernova 1987a was first discovered by Ian Shelton, working at Las Campanas Observatory in Chile*

△ Fig 12.25 *Before and after photographs of the supernova, with the progenitor star, Sanduleak −69° 202, arrowed*

Already we know which star exploded, a massive blue supergiant star a bit like Sirius, which had been catalogued by Nicholas Sanduleak in the 1960s. We also know that this supernova explosion is quite unlike any previously seen. For one thing it is the feeblest supernova explosion ever to merit the name. This at first caught the theoreticians on the hop, though they have been quick to realize that SN1987A's low optical luminosity was a result of the progenitor star having an unusually compact structure, due to the low abundance of heavy elements in the Large Magellanic Cloud.

The supernova has been studied at all available wavelengths – visible, infrared, ultraviolet (using the International Ultraviolet Explorer satellite which has been in orbit since 1978) and X-ray (using Japanese and Soviet satellites). It was also detected in a most unusual manner deep underground, both in America and Japan. Elementary particles called neutrinos, released when the star's core collapsed, were detected in giant underground laboratories at Cleveland, Ohio, and Kamioka, central Japan, at 07.35 GMT on 23 February. Neutrinos have no mass or charge, travel at the speed of light, and normally pass right through the earth, but a few can be detected by subtle means. Scientists in Italy and the Soviet Union also thought they had seen neutrinos from the supernova, but the timing of their events does not agree with those of the other experiments. In February 1989 a large consortium of astronomers announced that they had detected an optical pulsar in supernova 1987A, with the extraordinarily short pulsation period of $\frac{1}{2,000}$ of a second. This would be an exciting confirmation that a neutron star had formed from the collapsed stellar core.

Astronomers will continue to study the Large Magellanic Cloud supernova for years and decades to come, watching the development of the pulsar and following the motion of the remnant of the outer parts of the star.

Meanwhile let us leave the subject of supernovae with this thought. Many of the elements that make up the earth, especially those heavier than iron, were made in the interiors of massive stars like the one that Chinese astronomers saw explode in 1054. In those intertwining and delicate filaments of the Crab nebula, we see the material from which new solar systems will form, and ultimately, perhaps, new life.

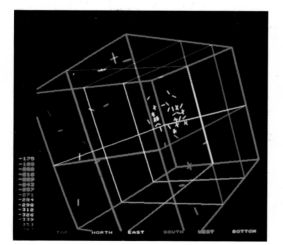

◁ *Fig 12.26 Detection of neutrinos from the supernova by the Irvine-Michigan-Brookhaven neutrino detector. This consists of a huge tank of purified water, monitored continuously by very sensitive photocells, which look for light flashes due to the passage of the neutrinos*

▷ *Fig 12.27 The Vela supernova remnant . . . from these filaments new solar systems will form, and perhaps new life . . .*

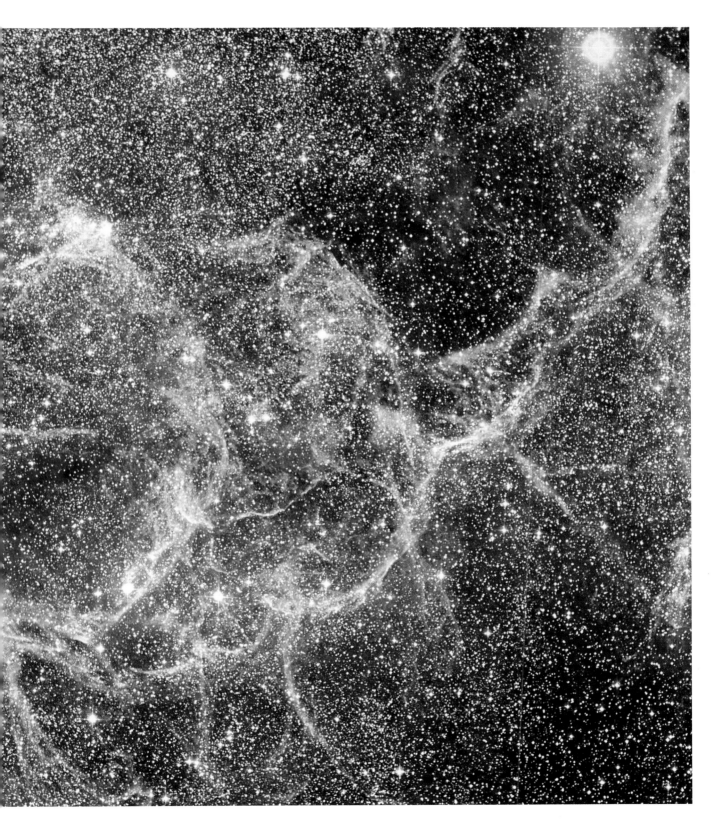

C H A P T E R   1 3

And eke the Bull hath with his bow-bent horne
So hardly butted those two twinnes of Jove,
That they have crusht the Crab, and quite him borne
Into the great Nemean lions grove

SPENSER *The Faerie Queen*

... the stars of the Swan and that other star, Aldebaran

GEORGE SEFERIS *'Epiphany'*

HYADES AND PLEIADES

STAR CLUSTERS

THE HYADES

The Hyades star cluster appears to the naked eye as a V-shaped group of stars defining the head of Taurus, the Bull. The eye of the Bull is the bright red-orange star Aldebaran, which is one of the most striking stars on northern winter evenings. It is also notable for being one of the few bright stars which can be eclipsed or 'occulted' by the moon. Studying records of an occultation of Aldebaran seen in Athens in March of AD 509, Edmund Halley concluded in 1718 that this could not have occurred unless Aldebaran had moved in the interim, the discovery of 'proper motion'. However Aldebaran is only a chance superposition on the Hyades cluster and does not belong to it.

◁ *Fig 13.1 The V-shaped Hyades cluster to the lower left, with the Pleiades cluster just above centre*

Taurus was one of the very earliest constellations to be recognized and was probably named as early as 4000 BC, when it marked the Spring Equinox and its meeting with the sun marked the beginning of the agricultural year. In later Greek myth, Taurus was identified with the snow-white Bull in whose guise Zeus carried off Europa. In another legend Taurus was the Cretan Bull, who hid in the depths of the labyrinth of Knossos.

The ancient Greeks and Romans associated the Hyades with wet and stormy weather. Homer refers to the 'rainy Hyades' and Pliny writes of them as

> . . . a star violent and troublesome;
> bringing forth storms and tempests
> raging both on land and sea . . .

It is interesting that the same tradition is found in ancient Chinese literature. In the *Shih Ching*, or *Book of Songs*, of the sixth century BC, we find

> The mountains and streams never end;
> The journey goes on and on . . .

> The Moon is caught in the Hyades;
> There will be great rains.
> The soldiers who are sent to the East
> Think only of this . . .

In Greek myth the Hyades were the daughters of Atlas and Aethra, half-sisters of the Pleiades. They were entrusted by Zeus with the care of the infant Bacchus and afterwards rewarded by a place in the heavens.

The Hyades cluster is one of the nearest groups of stars to the sun and holds a key place in the measurement of the size of the universe. Although not quite near enough to determine the distance to individual stars in the cluster by the method of parallax (chapter 2), statistical methods can be used to give a distance of 140 light years, with an uncertainty of 7 light years either way. This distance may soon be measured even more accurately from space, for example by the Hubble Space Telescope, which will be able to measure the parallax of individual stars in the cluster. The distances to other young star clusters resembling the Hyades can then be found by comparing the brightness of stars of the same colour and type in the two clusters. In this way the distances and luminosities of a few dozen Cepheid variable stars can be determined and these, through the period-luminosity relation (chapter 9), allow us to measure the distances to external galaxies. Just as the ancient constellation of Orion opens up a new world of stars being born out of clouds of gas between the stars, so this classical star-group, the Hyades, situated in one of the oldest of constellations, Taurus, the Bull, turns out to provide the benchmark for measuring the size of the universe.

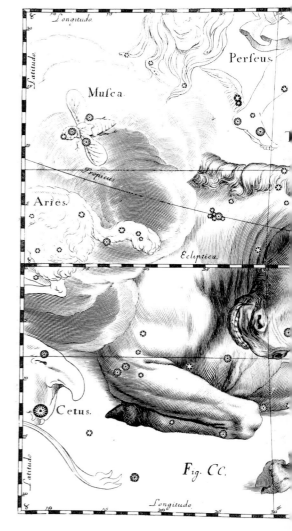

△ *Fig 13.2 The Taurus constellation according to Hevelius*

As when the seaman sees the Hyades

MARLOWE *Dr Faustus*

Thro' scudding drifts the rainy Hyades
Vext the dim sea

TENNYSON *'The Two Voices'*

THE PLEIADES

The Pleiades, the Seven Sisters, is the most famous of star clusters, the subject of myth since remote antiquity. The earliest recorded reference to them is in the Chinese annals, dated at 2357 BC, when they were close to the Spring Equinox. In Greek myth they were half-sisters of the Hyades and were saved by Zeus from pursuit by the giant Orion by being transformed into a group of celestial doves. Hesiod gives them a strong agricultural role:

When Atlas-born, the Pleiad stars arise
Before the Sun above the dawning skies,
'Tis time to reap, and when they sink below
The morn-illumined west, 'tis time to sow . . .

△ Fig 13.3 *The Pleiades star cluster*

◁ Fig 13.4 *Reflection nebulosity surrounding the Pleiades star Merope*

Now the Great Bear and Pleiades
where earth moves,
are drawing up the clouds
of human grief
breathing solemnity in the deep night.

BENJAMIN BRITTEN/MONTAGU SLATER *Peter Grimes*

while Virgil warned:

> Some that before the fall of the Pleiades
> Began to sow, deceived in the increase,
> Have reaped wild oats for wheat . . .

Later Manilius, in the days of Augustus, writes of them as the 'narrow cloudy trail of female stars'. The seventh century Arab poet Amr al Kais speaks of 'the hour when the Pleiades appeared in the firmament like the folds of a silken sash variously decked with gems'. The thirteenth century Persian poet Sadi wrote: 'The ground was as if strewn with pieces of enamel, and rows of Pleiades seemed to hang on the branches of the trees . . .' On a less elevated note, Sancho Panza, faithful servant of Don Quixote, visited them on his aerial voyage on Clavileno Aligero as the Seven Little Nanny Goats. The Onemdaga Iroquois Indians of North East America see them as the Seven Dancing Children.

For the Pre-Columbians of Central America, especially the Maya and the Aztecs, the Pleiades was the most important constellation in the sky. Its midnight culmination every fifty-two years, when their sacred and secular calendars once again came into coincidence, was a moment of particular dread and the signal for the most horrific human sacrifices. The west face of the Pyramid of the Sun at the ancient site of Teotihuacan, near Mexico City, and the city streets, are oriented towards the setting of the Pleiades. This midnight culmination was also the traditional date of the fearsome

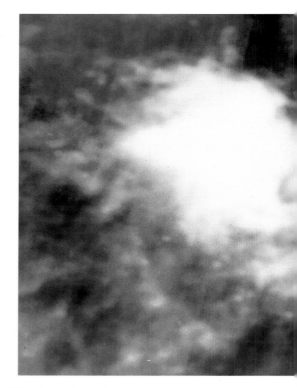

△ Fig 13.5 Infrared view of the region of the Pleiades cluster (left of centre), as seen by the IRAS satellite. In the bottom half of the picture can be seen the remains of the cloud of gas and dust from which the star cluster formed

◁ Fig 13.6 The double star cluster h and chi Persei, at a distance of 6200 light years. The hot blue stars show how young these clusters are, only a few million years old

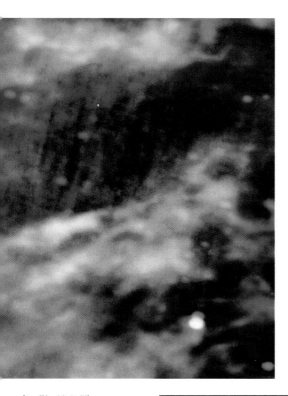

Witches' Sabbath or Walpurgisnacht, still observed today as Halloween. The traditional date of the culmination, 1 November, has been kept, though the midnight culmination of the Pleiades now occurs on 21 November. It is strange that both European and Meso-American cultures should attach such importance to the night when the Pleiades reaches its highest altitude at midnight.

To the average naked eye, the Pleiades cluster appears as a tight knot of six or seven stars, but some observers have recorded eleven or more under good conditions. Maistlin, the tutor of the sixteenth-century astronomer Johannes Kepler, is said to have been able to see fourteen before the telescope was discovered. With a large telescope, several hundred cluster members can be delineated. The cluster is about twenty million years old, which is very young by astronomical standards. The stars are still surrounded by wisps of gas, relics of the thick cloud of gas and dust from which the stars formed. Although twice as distant as the Hyades cluster, at three hundred light years, and beyond the range of present geometric distance methods, the Pleiades cluster also plays a valuable role in the measurement of cosmical distances. It is more similar to the distant clusters in which Cepheid variable stars are found than is the Hyades cluster.

The Hyades and Pleiades are examples of *open* or *Galactic* clusters, loose aggregates of stars formed relatively recently from clouds of gas and dust in the plane of the Milky Way. Much older are the 'globular' clusters, tight concentrations of hundreds of thousands of stars spread through the halo of the Milky Way.

▷ *Fig 13.7 The open cluster NGC3293*

The moon has gone down, the Pleiades have set. Night is half gone and life speeds by. I lie in bed, alone.

SAPPHO *'Light Vanishing'* c6 BC

The rainy Pleiads wester, Orion plunges prone . . .

HOUSMAN *More Poems XI*

OMEGA CENTAURI

To the naked eye Omega Centauri looks like a faint star and it was catalogued as such by Ptolemy in the first century AD. The name implies that it is the twenty-fourth brightest star in the constellation of Centaurus, the Centaur, for the brightest stars in each constellation were labelled in order of brightness – Alpha, Beta, Gamma, and so on down to the last letter of the Greek alphabet, Omega, by the German amateur astronomer Bayer in 1603. Modern astronomers still use his notation, extending it to a larger number of stars per constellation by using numbers (e.g. 61 Cygni, 30 Doradus).

In 1667, Omega Centauri swam into the field of view of William and Caroline Herschel's 25-foot telescope, with which they were systematically 'sweeping' or surveying the sky. They recognized at once that it was a rich and concentrated cluster of stars. Their son John, also a prominent astronomer, wrote of it: 'The noble cluster Omega Centauri is beyond all comparison the richest and largest of its kind in the heavens. The stars are literally innumerable . . .

Omega Centauri in fact contains about a million stars, crammed into a spherical star-cloud only sixty light years across. In the sun's locality there are only a few hundred stars in a volume this size. The condensed spherical shape gives the name *globular clusters* to such systems. They are the oldest structures in our Milky Way Galaxy, formed soon after its birth. As globular clusters age, the stars evolve with time and change their colour and brightness. First the most massive stars exhaust their hydrogen and become red supergiants and then supernovae. With time, less and less massive stars exhaust their nuclear fuels and die. Today, stars like the sun, in an older system like Omega Centauri, have already become red giants.

These characteristic evolutionary changes allow astronomers to estimate the ages of the globular clusters rather well. Since the time to form the clusters is believed to be quite short, the oldest globular clusters give an age for the whole Milky Way system itself. The best estimate of this age is thirteen thousand million years, but it might be as low as ten or as high as eighteen thousand million years. And our Galaxy is probably not much younger than the universe itself.

Omega Centauri is at the immense distance of twenty thousand light years away, and some globular clusters in the Milky Way are five times further than that. The globular clusters define a vast, roughly spherical halo around our Galaxy. They are the oldest structures we know of in the universe and they hold the key to understanding how our Galaxy and other galaxies formed.

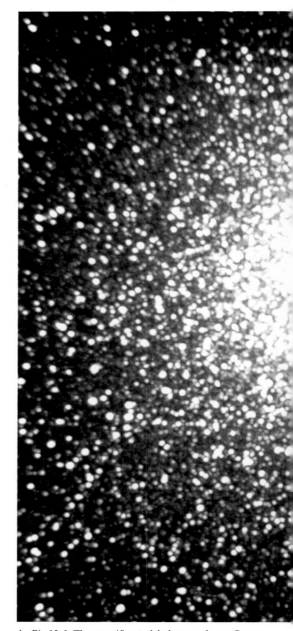

△ *Fig 13.8 The magnificent globular star cluster Omega Centauri*

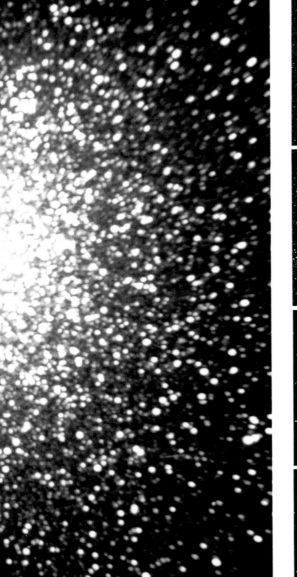

◁ Fig 13.9 Other
examples of globular
clusters:
a 47 Tucani
b Messier 19
c Messier 92
d Messier 3

C H A P T E R   1 4

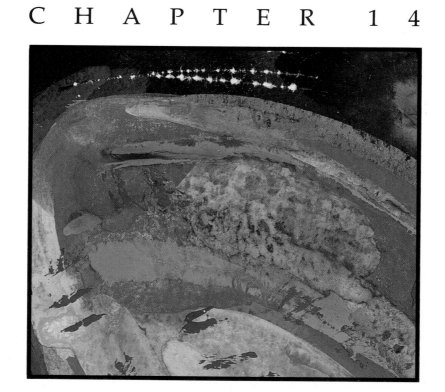

The Milky Way
revolves at night among the floating stars . . .

LI HO *'A Song of Heaven'* A D 791–817

THE MILKY WAY

OUR GALAXY

On a clear moonless night the Milky Way can be seen streaming across the sky. It is at its best from the southern hemisphere but still very striking on a northern summer night, and an especially marvellous sight from a high mountain site, such as the places astronomers like to build their telescopes. In the mythology of many ancient cultures the Milky Way is a heavenly River, a Sky-road, a Great Path to the world beyond, a cosmic bridge

◁ *Fig 14.1 Wide-angle view of Milky Way towards the centre of the Galaxy. The centre of our Galaxy is hidden from view at visible wavelengths by clouds of dust in the disc of the Galaxy along our line of sight*

linking Heaven and Earth. The splendour of the Milky Way is evoked in this poem from the ancient Chinese *Book of Songs*:

The men of the East, working endlessly,
But gaining no comfort;
The men of the West, splendid in their fine garments;
The sons of the boatmen,
Rudely clad in the skins of the bear;
The sons of the slave,
Taking whatever may be found;
If they have rich wine
They find no virtue in simple fare;
Their jade pendants are long
Yet they wish them longer . . .
While above
The Milky Way in Heaven
Shines on all brightly.

Almost two millennia later, the eleventh-century poet Su T'ung-Po muses:

It is nightfall, the clouds have vanished;
The sky is clear,
Pure and cold . . .
Silently I watch the River of Stars,
Turning in the Jade Vault . . .
Tonight I must enjoy life to the full,
For if I do not,
Next month, next year,
Who can know where I shall be?

In the first book of the *Metamorphoses*, Ovid writes: 'There is a way on high, conspicuous in the clear heavens, called the Milky Way, brilliant with its own brightness. By it the gods go to the dwelling of the great Thunderer and his royal abode.'

The concept of the Milky Way as a road used by the gods or immortals reappears in many human cultures, from the Norsemen to the American Indians and the Gabon pygmies. It has been a source of imagery for countless poets and in 1720 Jonathan Swift was moved to declare in his satirical *Edict* that aspiring authors and poets were forbidden to mention it. This did not deter Longfellow from writing *The Galaxy*, in which the Milky Way is ecstatically described as a

. . . torrent of light and river of the air,
Along whose bed the glimmering stars are seen
Like gold and silver sands in some ravine . . .

The nature of the Milky Way was speculated on by the Greek philosophers; Pythagoras and Democritus both appear to have believed it was composed of a vast number of faint stars. The Roman poet Manilius wrote at the beginning of the first century AD:

Is the spacious band serenely bright
From little stars, which there their beams unite,
And make one solid and continued light?

I climbed the ladder leaning against the hay,
into the uncombed loft.
I breathed the haydust of milky stars.
I breathed the matted scurf of space.

OSIP MANDELSTAM *Poems* (1928)

Shortly before, Diodorus of Sicily had noticed that the Milky Way formed a great circle around the sky (though he presumably had not seen the portion of the Milky Way that is visible only from the southern hemisphere), and suggested that it marked the line where the two starry hemispheres had been joined. Aristotle unfortunately adopted a rather vaguer position, that it was a 'gathering of celestial vapours' and it is not till Francis Bacon in the sixteenth century that we find again a more cogent view: 'The way of fortune is like the Milky Way in the sky, which is a meeting or knot of a number of small stars, not seen asunder, but giving light together. . . .'

It was Galileo who solved the problem of the nature of the Milky Way. In those exciting few weeks at the beginning of 1609 when he first turned his telescope on the sky, one of the most interesting of his discoveries was that the Milky Way became resolved into myriads of faint stars:

> I have observed the nature and material of the Milky Way. With the aid of the telescope this has been much scrutinized so directly and with such ocular certainty that all the disputes which have vexed philosophers through so many ages have been resolved, and we are at last freed from wordy debates about it. The Galaxy is, in fact, nothing but a congeries of innumerable stars grouped together in clusters.

△ Fig 14.2 Panorama of the Milky Way, dominated by the bright star clouds in Sagittarius close to the direction of the centre of the Galaxy. The constellation of Sagittarius fills the left half of the picture, the tail of Scorpio lies to the right of centre and Ophiuchus to the upper right

▷ Fig 14.3 Galileo, who in 1609 demonstrated that the Milky Way is composed of stars

The idea that the Milky Way is a vast disc-shaped aggregation of stars comprising all the stars seen by the naked eye or with a telescope was first put forward by Thomas Wright of Durham in 1750, as part of a strange and mystical conception of the universe. A more rational account of this theory was put forward by the German philosopher Immanuel Kant shortly afterwards. William and Caroline Herschel set out to map the structure of the Milky Way by counting the numbers of stars they could see in different directions. They concluded that the Milky Way had a 'grindstone' structure but their method, though ingenious, was in fact severely hampered by the dimming effect of the dust between the stars.

William Herschel at first believed that the nebulae, fuzzy patches of light that swam into his telescope's field of view in great numbers, were distant systems like the Milky Way. This 'island universe' theory had been advanced earlier by Christopher Wren, whose distinction as an astronomer was overshadowed by his achievements as an architect, and by Immanuel Kant. The fact that several objects which looked nebulous to a small telescope turned out to be star clusters when observed with a larger one supported this view. However Herschel's studies of planetary nebulae like the Ring Nebula (chapter 7) convinced him that those objects which still looked fuzzy in a large telescope were in fact shining clouds of gas from which new stars were forming, and at the end of his life he believed that the Milky Way comprised the whole universe. The gaseous interpretation of the nebulae appeared to find decisive support when in 1864 William Huggins showed with his spectroscope that the Orion Nebula was composed of hot gas.

There was only one really determined effort to revive the island universe theory in the nineteenth century. In 1845 William Parsons (Lord Rosse) completed an enormous reflecting telescope with a 72-inch mirror, the 'Leviathan', at his estate in Parsonstown, Ireland. With this he made some fascinating, and remarkably accurate, sketches of spiral structure in various nebulae, notably the famous spiral galaxy Messier 51. He also managed to resolve into stars some of the nebulae which Herschel had failed to resolve with his smaller telescopes. For a while Parsons' work revived interest in the idea that all, or at least most, nebulae were composed of stars and that the fainter, spiral nebulae were very distant Milky Way systems. The absence of any method of determining the distances of the nebulae, however, combined with the powerful evidence for the gaseous nature of nebulae like the Orion and Ring Nebulae, gradually undermined support for the theory. By the end of the nineteenth century the prevailing view was that the universe consisted solely of the Milky Way system.

The first twenty years of the twentieth century saw several major studies of the structure of the Milky Way. The Dutch astronomer Jacobus Kapteyn used star-counts to derive a model of the Milky Way in which the sun lay close to the centre of a huge disc. Like Herschel's earlier studies, this work was flawed by neglect of the role of interstellar dust, which Edward Barnard was beginning to uncover. Harlow Shapley used the globular clusters to locate the centre of our Galaxy, which he found to lie in the direction of Sagittarius. Of the star clouds in Sagittarius, Barnard wrote in 1913:

▷ Fig 14.4 William Parsons, Lord Rosse, who championed the island universe theory in the nineteenth century

▽ Fig 14.5 Rosse's 1850 sketch of the spiral galaxy Messier 51 (see Fig 16.6a for a modern photograph)

▽ Fig 14.6a Schematic view of Milky Way as seen edge-on

b Schematic view of Milky Way as it would appear from above, derived from radio observations of atomic hydrogen

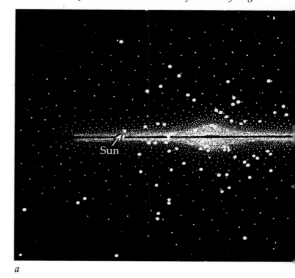

Sun

a

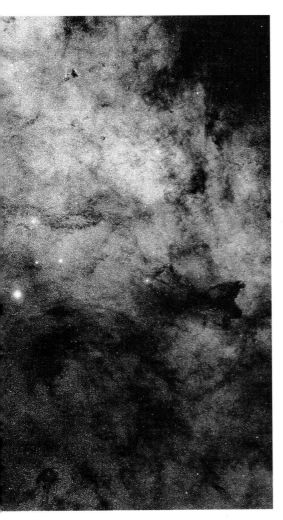

These magnificent star clouds are the finest in the sky. They are full of splendid details; one necessarily fails in an attempt to describe this wonderful region of star masses. They are like the billowy clouds of a summer afternoon; strong on the side toward the Sun, and melting away . . . on the other side. Forming abruptly at their western edge against a thinly star strewn space, these star clouds roll backwards toward the east in a broadening mass to fade away into the general sky . . .

In 1922 a dramatic debate was staged by the American Association for the Advancement of Science between Harlow Shapley and Heber Curtis. Shapley argued that the Milky Way included all known structures in the universe, including the spiral nebulae. Curtis advocated the island universe theory. The core of the debate was the issue of the size of the Milky Way system. Shapley, using his studies of globular clusters, arrived at a size about three times too great and concluded that the Magellanic Clouds therefore formed part of the Milky Way. Curtis, on the other hand, used Kapteyn's star counts to derive a size about three times too small. In both cases it was interstellar dust which caused the error, causing the star counts to fall away well before the edge of the Galaxy had been reached and making the globular clusters appear dimmer and further away than they really are. The debate about the size of our Galaxy is perhaps not over yet, for as recently as 1984 the agreed distance of the Galactic Centre from the sun was revised downwards by twenty per cent by the International Astronomical Union.

The picture we have today of our Galaxy is complex, and we are by no means sure of its past history. The basic structure is a *disc* of stars, gas and dust, with the gas and dust forming a thin layer inside the rather thicker stellar disc. Towards the centre of the disc the stellar distribution broadens to form an almost spherical *bulge* of stars. Within the thin disc of gas and dust are found the giant clouds of hydrogen, mostly in the form of molecules, out of which new stars are forming. Star formation appears to be concentrated in *spiral arms*, which trail out from the centre of the Galaxy. Although these are best observed in other galaxies, they have been painstakingly traced out in our Galaxy, especially at radio wavelengths.

The disc, which extends to a radius of about 50,000 light years (the sun is about half-way out from the centre) is surrounded by a roughly spherical *halo* of globular clusters about 200,000 light years in radius. These are older systems than most of the stars of the disc and are relics of the very earliest stages of the formation of our Galaxy. From the orbits of stars and gas clouds in the disc round our Galaxy, it is possible to deduce that there is a great deal of dark matter spread through the halo, perhaps comprising ninety per cent of the mass of the Galaxy. This could be in the form of Jupiter-sized objects or brown dwarfs, which would be too faint to detect, or, more fancifully, massive black holes a million times the mass of the sun. Both of these are just hypothetical explanations of how the halo could be massive without emitting detectable radiation at any wavelength. An even more exotic explanation is that there is some strange sub-atomic particle not yet discovered which is spread through the halo. Such particles have been given names which reflect their supposed role in particle physics, for example *photinos* or *axions*.

△ *Fig 14.7 Star clouds in Sagittarius*

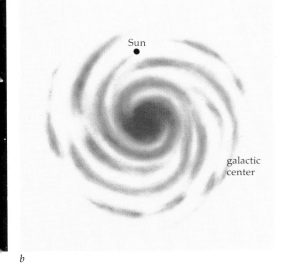

b

An important development in our understanding of the Galaxy came in 1933, when Karl Jansky detected radio waves from the Milky Way. He was working for Bell Telephone Labs on the problem of the hiss on transatlantic telephone lines. He built an antenna to try to locate the origin of this hiss and found to his surprise that the noise rotated with the stars and came from the direction of Sagittarius. He concluded that the hiss arose from the Milky Way. With this discovery began the modern astronomy of the invisible wavelengths.

After some further study of this emission, Jansky was then assigned by Bell Labs to more commercially profitable studies. Astronomers did not take a great interest in Jansky's discovery and it was left to an amateur, Grote Reber, to keep radio astronomy alive. He built a large steerable antenna in his back garden in Wheaton, Illinois, much to the curiosity of neighbours and passers-by, and used it to map the emission from the Milky Way in considerable detail. A story is told (by the late Bart Bok) that Reber submitted a paper on his work to the American *Astrophysical Journal*. Instead of sending the paper to another astronomer for comment, as is the usual practice, the editor, Otto Struve, sent a delegation to Wheaton to find out what was going on and to have a demonstration. Unfortunately they arrived on a Monday, and Reber could not move his antenna because his mother was using it as one end of her clothesline!

It was not till the wartime studies of John Hey in England that radio astronomy made a new start. On 12 February 1942 the German warships *Scharnhorst* and *Gneisnau* managed to slip down the English Channel

▷ *Fig 14.8 Karl Jansky with the antenna with which he discovered radio emission from the Milky Way in 1933*

▽ *Fig 14.9a The Milky Way viewed in molecular gas. The thin layer of cold dense gas from which new stars will form, mapped in microwave emission from the carbon monoxide molecule*

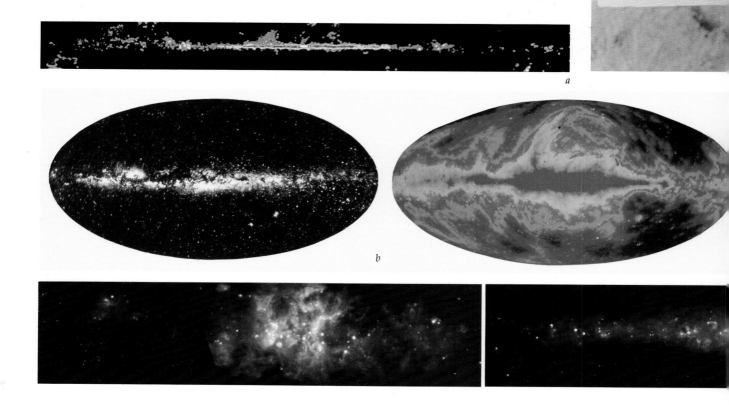

a

b

almost unnoticed because of German jamming of the British radar stations. Hey was given the task of examining the jamming menace. He set up a small group, based on the cliffs of Dover, to monitor the jamming. Then on 27 and 28 February there was a remarkably intense episode of radar jamming which knocked out all the coastal radar stations. Hey and his team realized that the direction of the jamming followed the sun, and when Hey telephoned the Royal Greenwich Observatory he learnt that an exceptionally active sunspot had crossed the solar disc on 28 February. Hey had discovered what G. C. Southworth had long been looking for at Bell Telephone Labs, radio emission from the sun. After the war, Hey's group went on to discover a strong discrete radio source in Cygnus which was to be of great significance for post-war astronomy (see chapter 17). Much later it was understood that Jansky's radio emission from the Milky Way was caused by *cosmic rays*, electrons moving at speeds close to that of light, spiralling through a large-scale magnetic field which permeates the Milky Way. Cosmic rays, which can be either electrons or atomic nuclei, were discovered early this century and most are believed to be accelerated in supernovae or pulsars.

▽ *Fig 14.9 Views of the Milky Way at different wavelengths:*
b Optical view. The general shape of the Galaxy can be seen, but extinction by dust blots out the central regions from view
c Radio view. The emission is from electrons moving close to the speed of light and the map essentially traces out the Galaxy's large-scale magnetic field
d Atomic hydrogen. A map in the characteristic 21-centimetre radiation of atomic hydrogen, showing the distribution of the gas between the stars
e X-rays. The emission comes from a local, low-density bubble of very hot (million degree Centigrade) gas left over from a nearby supernova explosion
f Infrared emission. Part of the Galactic Plane is displayed as a long strip. The brightest areas are due to dust warmed by newly formed hot stars

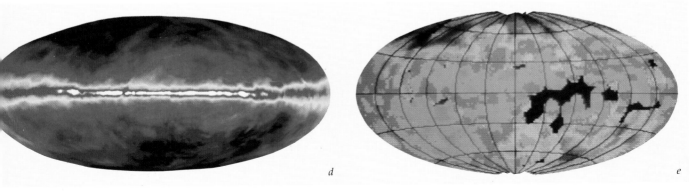

d

e

f

All the invisible wavelengths of modern astronomy are needed to study our Galaxy in its full richness. Atomic hydrogen can be mapped through its characteristic radio radiation at a wavelength of twenty-one centimetres. Molecular gas can be mapped at microwave wavelengths. Interstellar dust grains can be studied through the infrared radiation they emit. And at ultraviolet and X-ray wavelengths, we see the very hot gas left over from supernova explosions.

It is at radio, microwave and infrared wavelengths that we begin to see right to the heart of our Galaxy, a region known as the *nucleus* of the Galaxy, or the Galactic Centre. At optical wavelengths this is totally shrouded from view by the interstellar dust which lies in the plane of the Milky Way. The nucleus of our Galaxy has now been mapped in the radio emission which arises from electrons moving close to the speed of light, in the characteristic microwave light of carbon monoxide and many other molecules, in infrared emission from very hot and ionized atoms, and in the infrared emission from warm dust. From these studies we have gradually built up a picture of the dense zone within a few thousand light years of the centre of our Galaxy, though this picture is by no means final.

Firstly it is clear that in this central zone there is a rapidly rotating disc of molecular gas extending over several thousand light years. The star density is also very high here, though most were probably formed a considerable time ago. Our Galaxy is not one where exceptional star formation activity is taking place at the centre, like those I will describe in chapter 19. Near the centre of this disc, within a few hundred light years of the centre, we begin to see material which appears to be falling rapidly towards the centre. The actual centre appears to lie within the radio complex known as Sagittarius A, at the location of the very compact radio source Sagittarius A West. There is speculation that this could be a dormant black hole of mass about a million times the mass of the sun. Although we think of our Galaxy as relatively normal, we find that in its nucleus – visible to us only through the new astronomies of radio and infrared – it seems to harbour the same kind of monster as far more dramatic objects like the quasar 3C273, the subject of chapter 18. But at the moment the monster sleeps, waiting to be fed with gas and stars.

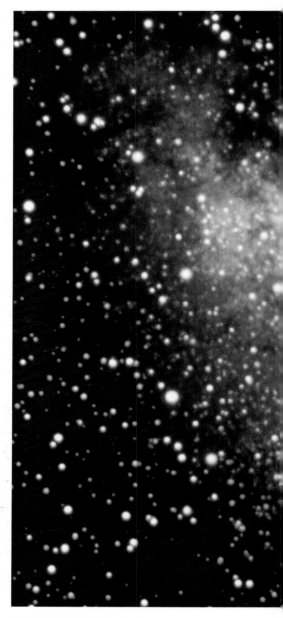

... the galaxy, that Milky Way
Which nightly as a circling zone thou seest
Powder'd with stars

MILTON *Paradise Lost*

For the earth would not have been a little
  light,
had not the Milky Way been there.
It and the stars.

KABBO *'The Girl of the Early Race who
        made the Stars'*
        (African Bushman)

▷ *Fig 14.10 The central regions of our Galaxy:*
a As seen at optical wavelengths. The centre lies in the obscured area to the right of the star cloud
b Far infrared view, colour-coded with cooler dust shown red, hotter dust shown blue
c Far infrared view, colour-coded with yellow as brighter emission, blue as fainter
d Near infrared view, showing the location of old giant stars
e Radio view of very central region. The centre of the Galaxy is close to the centre of the spiral-shaped feature, which is 3 light years across
f Radio view of central 200 light years of the Galaxy, showing strange plumes of material

b

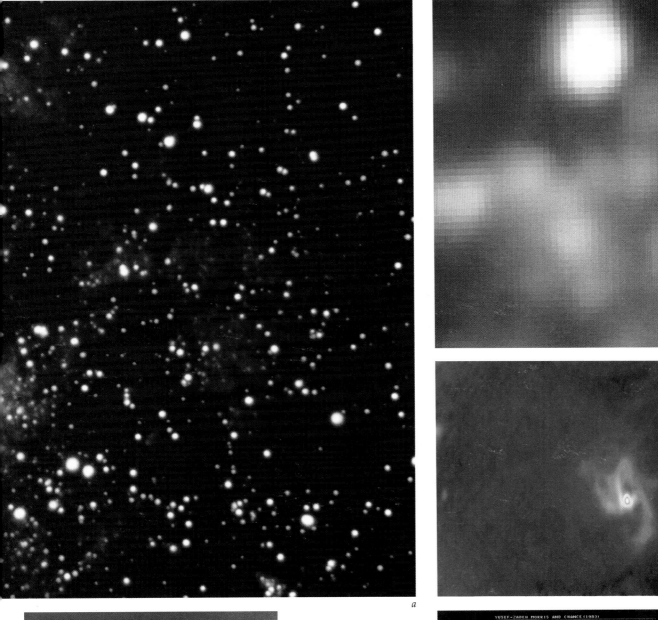

a

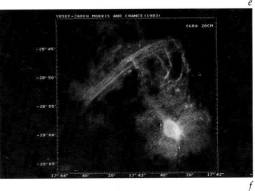

d

e

c

f

CHAPTER 15

I turned me to the right hand, and set my mind
on the other pole, and saw four stars
never yet seen save by the first people

DANTE *Purgatorio*

THE MAGELLANIC CLOUDS

OUR NEIGHBOURS IN THE UNIVERSE

The Large and Small Magellanic Clouds are visible to the naked eye in the southern sky and look like small detached portions of the Milky Way. They were first reported to Europe in 1520 by Pigafetta, chronicler of the 1492–3 circumnavigating expedition of the Portuguese mariner Ferdinand Magellan.

> The Antarctic Pole is not so starry as the Arctic. Many small stars clustered together are seen, which have the appearance of two clouds of mist. There is but little distance between them, and they are somewhat dim. In the midst of them are two large and not very luminous stars, which move only slightly. These two stars are the Antarctic Pole.

◁ *Fig 15.1 The Large and Small Magellanic Clouds, a spectacular sight to the naked eye in the southern hemisphere*

Magellan himself was killed during fighting with the inhabitants of Polynesia. The Clouds must have been familiar to the inhabitants of the southern hemisphere for millenia, and it is surprising perhaps that the Arab navigators, who must surely have crossed the equator centuries before Magellan, do not seem to have reported them back to their own sophisticated astronomers. It is possible that the object called Al-Bakr, the 'white ox', by Al-Sufi (903–986) is in fact the Large Magellanic Cloud. I remarked in chapter 4 that three thousand years ago the south celestial pole would have been near the Small Magellanic Cloud and this fact must surely have been valuable to the Polynesian navigators. Certainly the Clouds were used for navigational purposes by the European navigators of the sixteenth century.

In the early years of the twentieth century the Magellanic Clouds were still thought to be part of the Milky Way system, analogously perhaps to the Orion Nebula and its environs, another region of active star formation and nebulosity well outside the plane of the Milky Way.

The crucial step towards establishing the great distance of the Magellanic Clouds and their nature as external galaxies was the study by Henrietta Leavitt of Cepheid variable stars in the Clouds in 1908 (chapter 9). She noticed that these stars obey a 'period-luminosity law', the longer the period the greater the luminosity. Her results were published in 1912 but although she realized the potential of this law for measuring distance, she was not allowed to pursue this line of research by the director of the Observatory, Edward Pickering, who believed that the job of his staff was to collect data, not interpret it. Instead it was Harlow Shapley who calibrated the period-luminosity relation and used Cepheids to estimate the distances of the globular clusters in the halo of our Galaxy and of the Clouds. But as the size he obtained for the Milky Way system was about

▷ *Fig 15.2 At a distance of 150,000 light years, the Large Magellanic Cloud is the largest, and one of the nearest, of the satellite galaxies to our own Milky Way. In the picture the older, yellowish stars, the younger, hot blue stars and star clusters, and the reddish clouds of glowing interstellar gas excited by the most luminous stars of all, can be clearly distinguished*

15.3

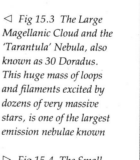

◁ *Fig 15.3 The Large Magellanic Cloud and the 'Tarantula' Nebula, also known as 30 Doradus. This huge mass of loops and filaments excited by dozens of very massive stars, is one of the largest emission nebulae known*

▷ *Fig 15.4 The Small Magellanic Cloud, which is at a distance of 200,000 light years. It is only about one sixth the mass of the Large Magellanic Cloud. Note the many young blue stars*

▷ *Fig 15.5 (far right) Deeper exposure of the Small Magellanic Cloud, in which a hint of spiral structure can be seen*

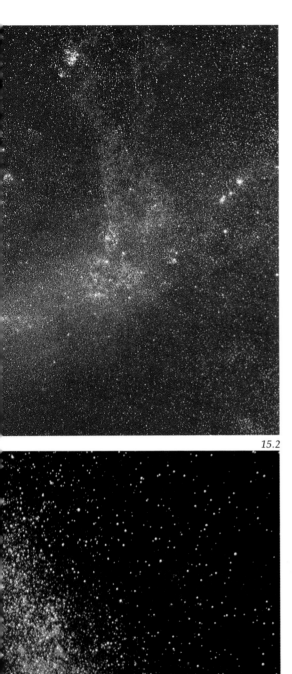

15.2

three times larger than it really is, he mistakenly thought the Clouds were part of our Galaxy.

Today we know that the Large Magellanic Cloud is about 150 thousand light years away, and the Small Magellanic Cloud about 200 thousand light years. This places them on the edge of the Milky Way system. They are quite small, irregularly shaped galaxies. The Large Cloud is about one quarter as luminous as our Galaxy and the Small Cloud about one twenty-fifth as luminous. Careful study shows that both probably have a dimly discernible spiral structure. Both galaxies are very rich in gas and are forming stars at a much slower rate than our Galaxy. They are bluer in colour than our Galaxy would be if we could see it from the outside because more of their light comes from recently formed stars. In our Galaxy much of the visible light, and most of the total output at all wavelengths, comes from older, redder stars. Another difference between the Magellanic Clouds and the disc of our Galaxy is that the abundance of heavy elements is much lower in the Clouds, reflecting their low star formation rates. To astronomers the 'heavy' elements are those from carbon, nitrogen, oxygen and onwards, which are made in the interior of stars.

As they orbit round our own much more massive Galaxy, the Magellanic Clouds interact tidally with it, in much the same way as the moon raises tides on the earth. Their gravitational attraction warps the outer parts of the disc of our Galaxy. Our Galaxy has a much more dramatic effect on the Clouds, drawing out a long stream of gas from the Clouds which plunges down towards the Milky Way, called the Magellanic Stream. The interaction may also play a part in generating the spiral arms in our Galaxy. It is noticeable that galaxies with companions tend to have much more pronounced and beautiful spiral arms. Astronomers call them 'Grand Design' spiral galaxies.

15.4

15.5

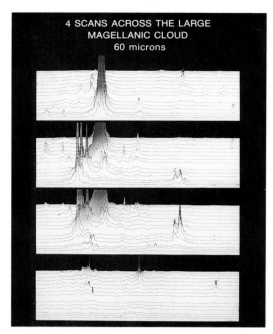

△ *Fig 15.6 Map of Tarantula Nebula region of the LMC, constructed by the author from observations made by the IRAS satellite during its first week of operation*

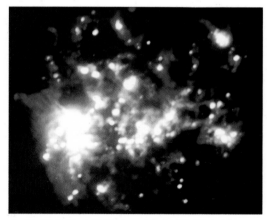

△ *Fig 15.7 Final IRAS map of the LMC. The bright source to the left is the Tarantula Nebula*

... but this attention must not make him forget even for a moment the explosion of a supernova taking place in the great Magellanic Cloud at the same instant, that is to say a few million years ago.

ITALO CALVINO
Mr Palomar

Within the Large Magellanic Cloud lies one of the largest star-forming clouds known, the Tarantula Nebula, 30 Doradi. Some astronomers have argued that it contains a star of over one thousand times the mass of the sun, ten times more massive than any star known in the Milky Way. It is more likely that the nebula is illuminated by a compact cluster of ten or twenty massive stars.

I came across this nebula in amusing circumstances. After the launch of the IRAS Infrared Astronomical Satellite in January 1983, there was a period of several days when we were still not exactly sure where the telescope was pointing. Tom Hibberd, a software engineer at Jet Propulsion Lab, Pasadena – the data analysis centre for the mission – and myself were checking out the computer programmes which were supposed to scan the raw data from the infrared detectors and pick out the astronomical sources. One afternoon, Tom brought me some plots of the detector output for an orbit from the previous day. For most of the orbit the detectors behaved normally, every so often responding to some astronomical source. Suddenly at one point all the detectors went haywire. 'What's this stuff, then?' Tom asked. Our first idea was that it must be some debris left over from the ejection of the satellite's cover a few days previously. However I stayed late that night, puzzling over the spaghetti-like jumble of signals. I found that the same thing had happened on several other orbits that day, and at about the same point on the orbit. Gradually I realized that the output from different orbits could be pieced together to make a map. The satellite was in fact passing over hundreds of very bright sources. When I went to Norton's *Star Atlas* to see roughly where the satellite was supposed to be pointing, the penny dropped. We were near the Large Magellanic Cloud and careful analysis of the satellite's pointing by another IRAS scientist, Eric Young, showed that the spaghetti was in fact the Tarantula Nebula. As they were in their usual difficulties about extracting funding from Congress, NASA were desperate to release some early results from IRAS and my cut-and-paste map was redrawn by JPL graphic artists in lurid blood-red colours for distribution to the world's media.

The Magellanic Clouds are special not only because they are our nearest neighbours in the vast universe of galaxies and the most prominent of our small gaggle of faithful dwarf companions. They are also an important testing ground for methods of measuring the distances to galaxies, a matter of great controversy for over sixty years. The explosion of supernova 1987A near the Tarantula Nebula was important not only for giving us a wonderful opportunity to study the mechanism of a supernova explosion at relatively close (but not too close!) quarters, it also allowed us to check that supernovae can indeed be used to measure galaxy distances. As we have seen, supernova 1987A was rather unusual and caught astronomers by surprise. But when they thought about it, they soon realized they could explain it and they then found that the distance of the supernova agrees very well with the distance to the Large Magellanic Cloud determined from Cepheid variable stars, novae and other distance indicators. To measure the distance to the galaxies seems to me to be one of the greatest achievements of human beings, with nothing but their powers of observation and their ingenuity to free them from the bounds of this insignificant speck of dust, the earth.

△ *Fig 15.8 The 30 Doradus (Tarantula) Nebula in the LMC. The filaments extend to over a thousand light years, making it 30 times larger than the Orion Nebula in our Galaxy*

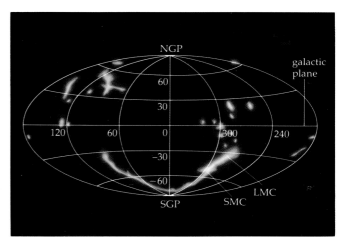

△ *Fig 15.9 The Magellanic Stream, a ribbon of gas which spans the gulf between our Galaxy and its two satellites, the Large and Small Magellanic Clouds (LMC and SMC). It is believed to have been torn off the Clouds by our Galaxy's gravitational attraction*

△ *Fig 15.10 Close-up of the Tarantula*

C H A P T E R   1 6

I see beyond this island universe,
Beyond our sun, and all those other suns
That throng the Milky Way, far, far beyond,
A thousand little wisps, faint nebulae,
. . .
Faint as the mist by one dewdrop breathed
At dawn, and yet a universe like our own;
Each wisp a universe, a vast galaxy
Wide as our night of stars.

ALFRED NOYES *The Torch Bearers*

THE ANDROMEDA NEBULA

TWIN TO OUR GALAXY

Visible to the naked eye as a small fuzzy object in the constellation of Andromeda, the Andromeda Nebula was first noticed by Arab astronomers of the tenth century AD. In his *Book of the Stars* of AD 905, Al Sufi recorded the nebula as 'a little cloud'. Simon Marius appears to have observed it through a telescope in 1611 or 1612, only a couple of years after Galileo's invention of the astronomical telescope. He compared the soft glow to 'the light of a candle shining through horn'. Messier included the nebula as number 31 in his Catalogue of Nebulous Objects of 1784 (p. 51).

Though some astronomers of the subsequent centuries thought the

◁ *Fig 16.1 The constellation of Andromeda, with the Andromeda Nebula, Messier 31, just below the centre of the picture. Another nearby galaxy, Messier 32, can be seen just below M31*

Andromeda Nebula to be a cloud of gas from which a new solar system was forming, others like Christopher Wren dared to speculate that it was an 'island universe' comparable to our Milky Way system. As we have seen, studies of other types of nebulae in the eighteenth and nineteenth century cast doubts on this grand idea.

On 20 August 1885, E. Hartwig discovered with a telescope a new star or nova near the centre of the Andromeda Nebula which was so bright as to be almost visible to the naked eye. The star decreased in brightness over the subsequent months and was last observed, ten thousand times weaker, on 1 February 1886. If this 'nova' were similar to other novae in the Milky Way, the distance of the Andromeda Nebula would not be very great and it would have to be part of the Milky Way system. It was only in the 1930s, when Fritz Zwicky recognized that there must be two types of novae – the classical *novae* (chapter 11) and the far more luminous *supernovae* (chapter 12), which he attributed to the explosion of a massive star – that the nature of the event of 1885 became clear. It was in fact a Type I supernova, the earliest recorded outside our Galaxy.

The great distance to the Andromeda Nebula was not established till 1923, when Edwin Hubble was able to detect several Cepheid variable stars in the galaxy with the 100-inch telescope on Mount Wilson in California. Using Henrietta Leavitt's period-luminosity law (p. 71) as calibrated by Shapley in 1916, Hubble announced a distance of 900,000 light years, far beyond the edge of the Milky Way system. When Hubble's paper was read to the American Association for the Advancement of Science in December 1924, the audience knew that the island-universe debate was over. The first steps into the universe of galaxies had been taken.

In 1953, Walter Baade realized that the Cepheids used by Hubble were more luminous than the latter had thought, and the distance was revised

a

b

△ *Fig 16.2a Detailed view of M31, showing the spiral arms and dark dust lanes between them*
b Enlargement of the central region of M31

◁ *Fig 16.3 Close-up of Andromeda Nebula. Its two dwarf elliptical companions can also be seen*

to 2.0 million light years, close to the value accepted today (2.2 million light years). There are in fact two types of Cepheids: the first type are found amongst the young stars in the discs of spiral galaxies. The time for the variations in their brightness ranges from a few to a few hundred days. These are the ones which can be detected in external galaxies. The second type are much older stars found mainly in the halo of our Galaxy (the star Polaris is also one). The time for the variations in their brightness ranges from a few hours to a few days. For a Cepheid with a variability period of a few days, the first type of Cepheid is much more luminous than the second type. Unfortunately the stars used by Shapley to calibrate the period-luminosity law were of the second type, hence Hubble's underestimation of the distance to Andromeda.

The Messier 31 or M31 galaxy, as the Andromeda Nebula is known by astronomers, is close to being a twin to our Galaxy. It contains several hundred thousand million individual stars and its mass is at least one hundred thousand million times the mass of the sun. The spiral arms which can be seen trailing out from the central regions of the galaxy, or nucleus, are the location of regions of energetic formation of new stars.

It was in the Andromeda Nebula that Walter Baade first discovered the existence of two distinct stellar 'populations', using the Mount Wilson 100-inch telescope during the exceptional conditions of a Los Angeles wartime blackout in 1944. The stars in the spiral arms of the galaxy were blue and luminous like those found in open clusters in our Galaxy, such as the Hyades and Pleiades (chapter 13). These formed Baade's Population I and are relatively young stars. On the other hand, the stars in the nucleus of the galaxy resembled those in the globular clusters of our Galaxy. These Baade called Population II and he realized that they were much older stars.

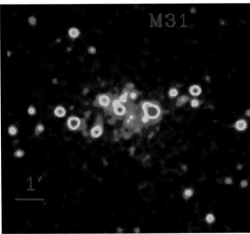

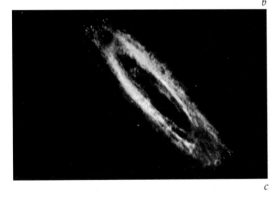

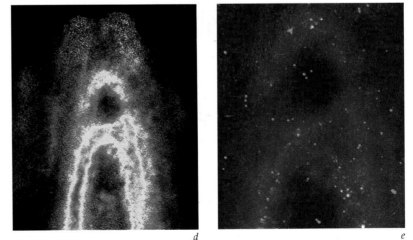

△ *Fig 16.4a Far infrared view of M31, showing where star formation is concentrated today*
b X-ray picture of the centre of M31. Most of the individual sources are X-ray emitting binary star systems
c Radio picture of hydrogen in M31, colour-coded to show the material coming towards us (blue) and moving away from us (red), so that the rotation of the galaxy can be seen

d Map of radio emission from north-east end of M31, colour-coded by intensity (red=bright, blue=faint)
e Composite picture of north-east end of M31, with emission from atomic hydrogen shown blue, radio continuous emission shown green, and optical emission from ionized hydrogen (H-alpha) shown red. The red or white spots are regions where massive new stars have formed recently and the green spots are supernova remnants

△ Fig 16.5 Other spiral galaxies:
a NGC 2997
b Messier 83
c NGC 4027

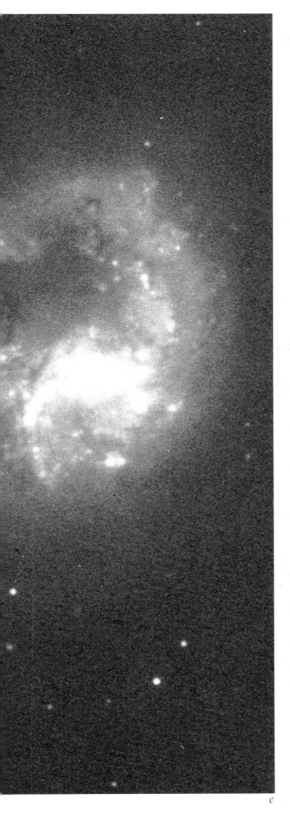

d Messier 33, a member of the Local Group of galaxies
e The 'Sombrero' galaxy, a spiral galaxy seen almost edge-on. The dark band across the
centre of the galaxy is the gas and dust of the galaxy's disc

c

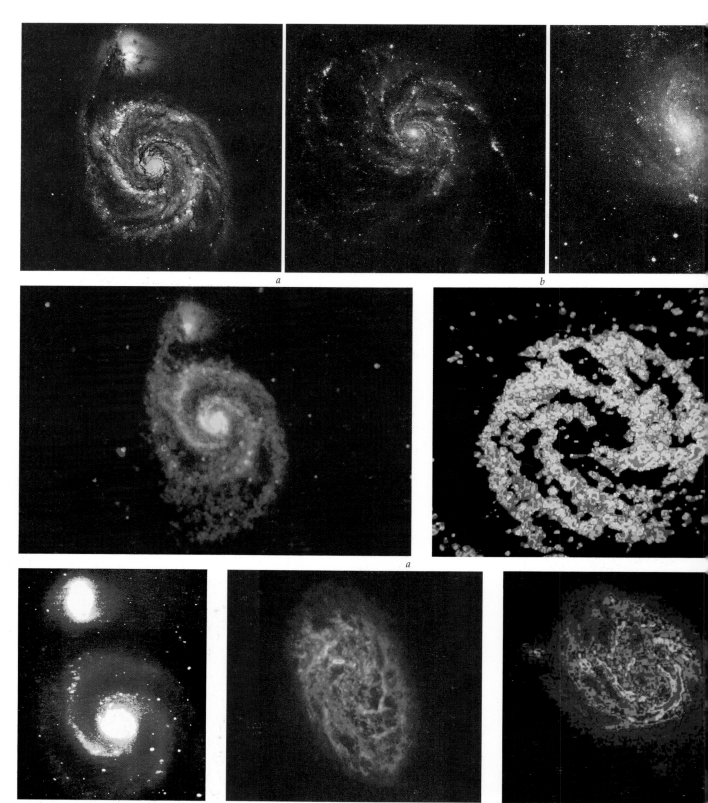

a

b

a

c

d

c   d

△ Fig 16.6 More spiral galaxies
a The 'Whirlpool' galaxy, Messier 51. See p. 112 for
William Parsons' 1850 sketch of this
b Messier 101
c NGC300
d NGC 1097, an example of a 'barred' spiral, with the
spiral arms starting from the ends of a bar

◁ Fig 16.7a Composite picture of Messier 51, with optical
light (mainly young stars) shown green, radio emission
from atomic hydrogen shown blue and radio continuous
emission shown red
b Radio picture of the atomic hydrogen in Messier 83, with
the directions of strongest emission shown red
c Near infrared image of Messier 51 and its companion, in
which stars and ionized gas clouds are seen unobscured by
dust
d Radio picture of the atomic hydrogen in Messier 33, one
of the galaxies of the Local Group
e Radio view of Messier 101

From these and subsequent studies, a picture of the formation and evolution of spiral galaxies like the Milky Way and the Andromeda galaxy has emerged. Before any stars formed, the galaxy would be an extended cloud of gas, much larger than it is today, falling together under gravity. During this collapse, the globular clusters, the oldest structures in the galaxy, must have condensed out into their tight aggregations of millions of stars. These would form a halo around the galaxy, with individual clusters ever plunging inwards and outwards in a memory of the original collapse. Meanwhile the remaining gas from the original cloud continued its collapse. This would gradually spin faster because the original cloud would have had some rotation or because the tidal attraction of other nearby protogalaxies would make it rotate faster. The cloud collapsed to form a rotating disc, out of which stars started to form. From then onwards the galaxy would look much as it does now, with stars continuing to condense out of the disc of gas. The actual mechanism which causes clouds of gas to start forming stars today appears to be a spiral compression wave, driven by the gravitational attraction of the stars, which travels round the galaxy, generating the elegant spiral arms seen in the Andromeda Nebula and painstakingly traced out in our Galaxy.

This basic picture of how galaxies are born is, however, only one possibility. Astronomers have grappled for decades with the problem of how galaxies form and we have by no means arrived at the final answer. In some theories the period of formation is a prolonged process of aggregation, with smaller pieces merging together over billions of years before the gas collapses to form the disc. In others the disc is built up gradually from a rain of gas falling from the halo of the galaxy. And in still other theories, a crucial role is played by *dark matter*, matter which is detectable only through its gravitational effect, perhaps in the form of as yet undiscovered exotic sub-atomic particles, the visible matter merely aggregating to pre-existing concentrations of dark matter.

The Andromeda Nebula and our Galaxy are the dominant members of a group of twenty or so galaxies known as the Local Group of Galaxies, which move through the universe together. Most of these galaxies are extremely small, even compared with the Magellanic Clouds, and are satellites either of our Galaxy or Andromeda. If we think of our Galaxy as a city, then these companions are just the outlying villages. The Local Group is a minor republic in the world of galaxies.

▷ Fig 16.8 The Local Group of Galaxies, dominated by
our Galaxy and the Andromeda Nebula, as it might appear
from near our Galaxy. Each of these giant spirals is
surrounded by several small satellite galaxies

C H A P T E R   1 7

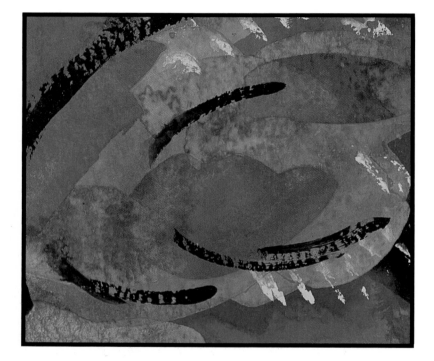

And from above thin squeaks of radio static,
The captured fume of space, foams in our ears . . .

HART CRANE *The Bridge*

MESSIER 87

RADIO GALAXY

Situated in the constellation of Virgo, Messier 87 is an example of a massive *elliptical* galaxy, an almost spherical distribution of over a million million stars. Charles Messier discovered it in March 1781 and described it as 'Nebula without star in Virgo . . . appears to have the same light as the two nebulae M84 and M86 . . .'. M87 is the dominant galaxy of the Virgo cluster of galaxies (chapter 20), of which M84 and M86 are also bright members, at a distance of about sixty million light years. Only one-fortieth times as bright as the faintest star visible to the naked eye, M87 needs at least a six-inch telescope or good binoculars to see it clearly. Surrounding the main stellar distribution in M87 is a remarkable cloud of thousands of globular clusters.

◁ *Fig 17.1 Optical picture of the giant elliptical galaxy, Messier 87, which is surrounded by a vast halo of globular clusters*

△ Fig 17.2a Close-up of NGC205, dwarf elliptical
companion of M31
b Two large southern ellipticals
c A lenticular galaxy

a

△ *Fig 17.3a Short
exposure optical
photograph showing the
M87 jet
b Detailed view of M87
radio source, showing the
jet broken up into a series
of knots of emission and
the two extended lobes of
emission on either side of
the galaxy. The nucleus of
the galaxy is the bright
spot at the end of the jet
and the jet is 4000 light
years long
c Close-up of the M87
nuclear jet made with a
network of radio-telescopes
spread around the world.
The inner 60 light years of
the jet is shown
d The M87 jet as seen in
X-rays*

Much smaller examples of elliptical galaxies are the dwarf ellipticals of the Local Group, for example the two dwarf companions of the Andromeda Nebula, which were first discovered by Caroline Herschel. Elliptical galaxies differ from spirals in several ways: they have little gas or dust and little or no current star formation going on. They are roughly spherical aggregates of old stars and look as if most of their stars formed in a single event many thousands of millions of years ago. Their redder colours reflect the older stellar population of which they are composed.

Intermediate between elliptical and spiral galaxies are *lenticular* galaxies which, in addition to the main ellipsoidal distribution of stars, have a disc of old stars with little associated gas or new star formation. One possible explanation is that they are spiral galaxies which have been swept clean of gas and dust, perhaps as a result of interaction with other galaxies in a close encounter. Most galaxies can be classified as elliptical, lenticular, spiral or irregular (like the Magellanic Clouds). About one in ten galaxies do not fit very well into these categories and are classified as *peculiar*.

In 1918, Heber Curtis at the Lick Observatory noticed in the central regions of M87: 'A curious straight ray lies in a gap in the nebulosity . . . apparently connected with the nucleus by a thin line of matter . . .'. This nebulous 'jet' of emission attracted little interest for thirty years. In 1948 John Bolton and his colleagues in Australia, following up the remarkable discovery by John Hey and his group of a powerful point source of radio emission in Cygnus two years earlier, discovered another strong source, this time in the constellation of Virgo. In the following year they were able to measure an improved position for the radio source and found that it was associated with Messier 87.

For several years, however, it was still assumed that most of the radio sources were some kind of dark star. A few more daring and speculative spirits like Tommy Gold, Fred Hoyle, Herman Bondi and Cyril Hazard argued that they were distant and powerfully radio-emitting galaxies. In a letter written in September 1951, the American optical astronomer Walter Baade wrote:

> The radio source in Virgo coincides with one of the brightest galaxies of the Virgo cluster of nebulae, NGC 4486 (M87). This coincidence could, of course, be accidental. But M87, a giant ellipsoidal galaxy, is unique among all the galaxies of its kind on account of a most unusual feature: a huge, thin jet of matter emanating from its nucleus. The nature of this jet is a mystery at present. Nevertheless, one begins to wonder whether the coincidence of the radio source and M87 is merely accidental.

Less than a year later, Baade found that Hey's strong source, which had been called Cygnus A, was associated with a distant elliptical galaxy with a curious hour-glass shape, the brightest member of a cluster of galaxies. This was the turning-point. Soon there were many examples of *radio-galaxies* and it gradually came to be realized that most of the radio sources were in fact far outside our Galaxy. Baade and Minkowski's dramatic theory for the origin of the huge amount of energy in the radio-source Cygnus A was that it resulted from a direct collision between two galaxies. This idea went out of fashion during the period 1965–85, but has suddenly revived again as the explanation of very luminous infrared galaxies (chapter 19). As more and more radio-galaxies were identified during the

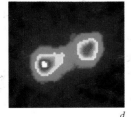

d

1950s and early 1960s, it became apparent that most of them were identified with single galaxies rather than colliding galaxies. The idea that emerged in the 1960s was that radio-galaxies were the result of explosions in the nuclei of galaxies. This theory was strongly advocated by the Armenian astronomer Victor Ambartsumian, and seemed to be proved by a classic study of the galaxy Messier 82 by Roger Lynds and Allan Sandage (see chapter 19). This explosion was supposed to result from the gravitational collapse of a huge mass of one hundred million solar masses of gas or stars. The mechanism for the production of the radio waves had been identified by the Russian physicist I. S. Shklovsky as due to the spiralling of electrons moving close to the speed of light in a strong magnetic field. This phenomenon had been observed in terrestial particle accelerators known as 'synchrotrons', so it became known as *synchrotron radiation*. The particles moving close to the speed of light are called *relativistic particles*.

As more and more powerful radio telescopes began to be built and extragalactic radio sources were mapped with ever improving resolution, a new picture began to emerge, less violent but no less fascinating. Narrow jets centred on galactic nuclei like that in Messier 87 turned out to be the rule rather than the exception, and it was clear that these jets, often extending on both sides of the galaxy, delineated the channel through which energy was being poured into the radio sources.

◁ *Fig 17.4a (left and below) The radio-galaxy Cygnus A at optical wavelengths. Baade and Minkowski interpreted this as a collision between two galaxies*
b Radio map of Cygnus A. The two lobes of emission, 300,000 light years apart, are powered by narrow twin beams emanating from the galaxy at the centre
c False-colour map of the Cygnus A radio-source
d Cygnus A in X-rays. The radio-lobes lie within the central spot. Most of the emission comes from very hot gas

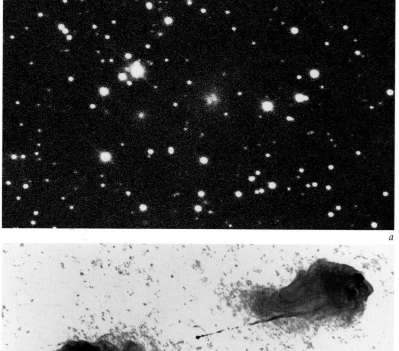

a

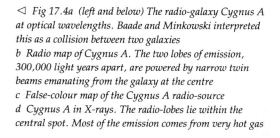

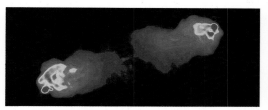

c

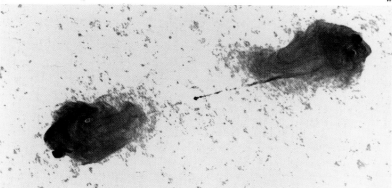

b

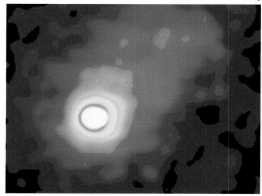

d

The jet in M87 is now believed to be due to a beam of particles and energy moving very close to the speed of light directed out of the nucleus of the galaxy. The engine which drives this beam is, according to the theory for radio-galaxies put forward by Roger Blandford and Martin Rees in 1974, a gigantic black hole whose mass is at least a hundred million times that of the sun. Intriguing indirect evidence for this has been seen in the form of its effect on the distribution of stars in the centre of the galaxy, which peak up to a remarkably high density and show evidence of very rapid motions, as would be expected if there were a mass concentration at the centre.

In 1977 Halton Arp and Jean Lorre used computer processing of high-quality images of the optical jet to show that it is in fact broken up into a series of bright knots of emission. These are believed to be due to the impact of the beam of relativistic particles defining the radio jet on gas trapped within it. The same structure has now been mapped in radio waves using the Very Large Array radio telescope in New Mexico, USA. This consists of twenty-five antennae arranged in the form of a Y twenty-five miles long, with which astronomers can map radio sources in extremely fine detail.

The radio structure of Cygnus A shows a double beam of emission emerging from the nucleus of the galaxy and expanding into two enormous lobes of emission located millions of light years from the parent galaxy. These beams, which preserve their direction over millions of light years, are a feature of many radio galaxies. The beam of relativistic particles impinges on gas clouds surrounding the galaxy and re-accelerate particles to speeds near the velocity of light which expand out into the lobes, radiating as they move through the accumulated magnetic field. The lobes are like a cloud of dust thrown up by a high-pressure air jet aimed at dusty ground.

a

△ *Fig 17.5a Optical photo of radio-galaxy Centaurus A, an elliptical galaxy with an unusual dust line, perhaps the result of a merger between an elliptical and a spiral*
b (below) Radio picture of the inner lobes of Centaurus A, the closest powerful radio galaxy. The full extent of the radio source is twenty times greater than shown here
c The nucleus of Centaurus A in X-rays
d Optical photo of Cen A with radio contours superposed

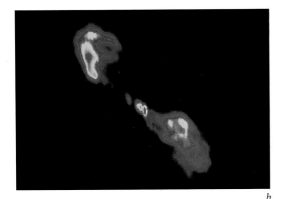

b

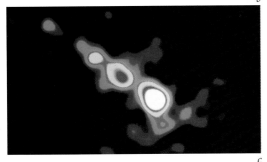

c

d

Radio-galaxies show a wonderful variety of forms. Sometimes the central engine appears to be being whirled around and the jets spray out at an ever-changing angle. Other galaxies are ploughing through the intergalactic gas and the jets are dragged back behind the galaxy in a wake. Yet the basic mechanism which powers the radio-sources always seems to be similar. Naturally we would like to be surer about this massive black hole which has to be postulated at the heart of the machine. But how can we ever be sure about something which is by definition invisible?

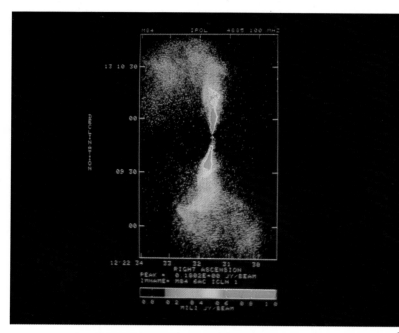

a

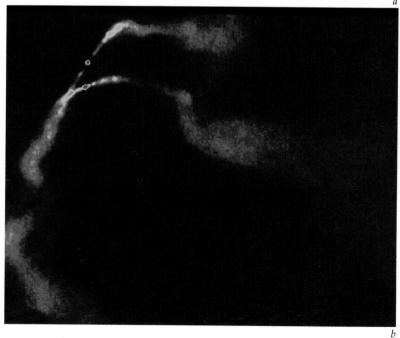

b

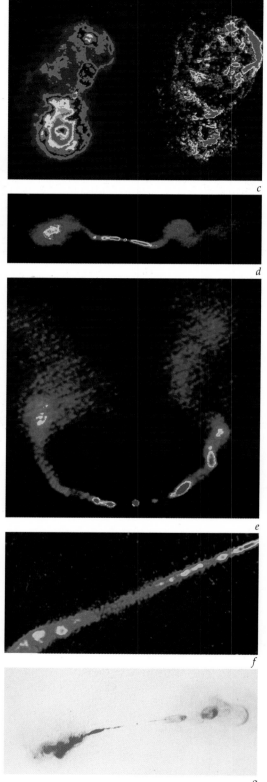

c

d

e

f

g

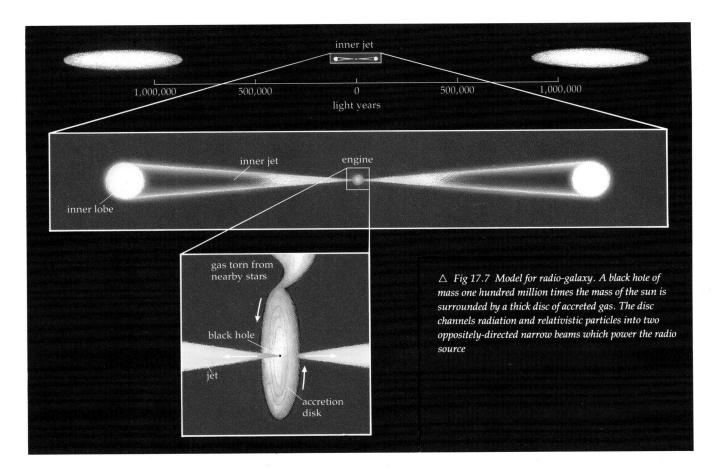

inner jet

1,000,000   500,000   0   500,000   1,000,000

light years

inner jet   engine

inner lobe

gas torn from
nearby stars

black hole

jet

accretion
disk

△ Fig 17.7 Model for radio-galaxy. A black hole of
mass one hundred million times the mass of the sun is
surrounded by a thick disc of accreted gas. The disc
channels radiation and relativistic particles into two
oppositely-directed narrow beams which power the radio
source

◁ Fig 17.6 Radio maps of galaxies:
a Messier 84. Two-sided jets and lobes of a weak radio galaxy
b 3C75. There are two bright centres of emission about
20,000 light years apart, each the source of twin radio jets
which appear twisted as they splay out into space. The
whole source is about a million light-years across
c 3C310, which appears to be blowing bubbles. The bright
red spot is centred on an elliptical galaxy whose size is about
the same as the radio spot. The right-hand image, in
polarised radio waves, brings out the filamentary shells
d 3C449, dual radio jets emerging from opposite sides of a
giant elliptical galaxy
e NGC1265. The twin radio jets are being swept backwards
as the galaxy moves through the intergalactic medium of the
Perseus cluster of galaxies (see chapter 20)
f NGC6251. This radio jet is over 300,000 light years long
and can be seen to be slowly spreading out as it expands
away from the galaxy at the lower left-hand end
g Hercules A. Two jets streaking away from the small faint
nucleus reach half a million light years into space. The
lower jet shows wiggles indicative of an unstable flow while
the upper jet shows several distinct rings each larger than
our entire Milky Way Galaxy

△ Fig 17.8 The Very Large Array radio telescope at Socorro, New Mexico USA. It
consists of 27 identical 25-metre antennae arranged along a giant Y twenty-five miles across

C H A P T E R 1 8

There is a remote possibility that it may be a very distant
galaxy of stars . . .

ALLAN SANDAGE *on 3C48, December 1960*

3 C 2 7 3

E N I G M A T I C   Q U A S A R

The story of the discovery of *quasars* and the subsequent controversy that they generated is a fascinating one. Following the post-war discoveries by Hey, Bolton and others of bright radio sources, Martin Ryle and his radio group at Cambridge set out to make a catalogue of point-like radio sources in the northern sky. At the same time, John Bolton and his colleagues at the Parkes Radio Observatory in Australia began to do the same in the southern hemisphere. When this work was begun, in the 1950s, it was still completely unclear whether most of these radio sources were unusual stars in our Galaxy or distant galaxies.

After a period of controversy between the two groups, from which it transpired that the Cambridge group had not been quite careful enough about what should be called a radio source, the Cambridge radio-astronomers produced in 1959 the profoundly influential *3rd Cambridge Catalogue of Radio Sources* (3C), a list of some five hundred reliable sources of radio emission. Although the positions of these sources were not very accurately known, some of them could already be identified: 3C144 was the Crab Nebula, 3C231 was the Andromeda Nebula and 3C274 was M87, for example. In 1960 Allan Sandage announced at the December meeting of American Astronomical Society in New York that the source 3C48 was identified with 'a 16th magnitude object in Triangulum that appears to be the first case where strong radio emission originates from an optically observed star'. The optical spectrum showed 'a combination of absorption and emission lines unlike that of any other star known . . . There is a remote possibility that it may be a very distant galaxy of stars: but there is general agreement among the astronomers concerned that it is a relatively nearby star with most peculiar properties.' By 1963 two more examples of radio stars had been found by Sandage and Thomas Matthews.

The crucial step forward occurred when, in the same year, Cyril Hazard and his colleagues in Sydney, Australia, managed to measure a very

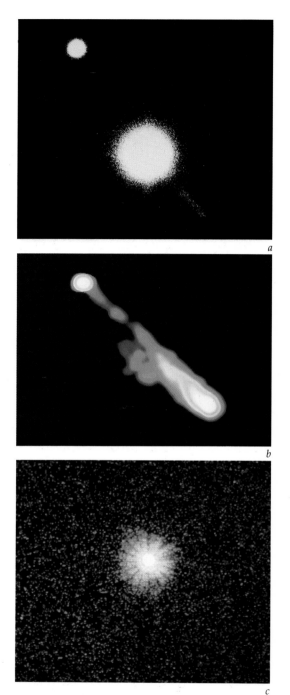

△ *Fig 18.1a The quasar 3C273. The jet can be seen faintly to the lower right*

b Radio picture of 3C273 made with the UK Merlin array of radio-telescopes, showing a compact nuclear source (upper left) and a jet extending out on one side to a distance of 150,000 light years

c 3C273 in X-rays, showing only the nuclear source

accurate position for the radio source 3C273 when it was occulted by the moon. By timing the exact moment when the radio source disappeared and reappeared, and using the known position of the moon, they could reconstruct the position of the radio source to the unprecedented accuracy of one second of arc (one 3600th of a degree). They immediately noticed that this position coincided with an unusual looking star, which had a faint wisp or jet sticking out from it. Although this 'star' is much too faint to be seen with the naked eye or binoculars, it is quite bright when seen in a large telescope or on the Sky Survey photographs published by the Mount Palomar Observatory.

The position of 3C273 was sent to Maartin Schmidt at Mount Palomar, who rapidly measured the a spectrum of this unusual star. He puzzled over the strange pattern of lines seen across the spectrum. Then he noticed that if he assumed the lines were simply emission lines of hydrogen but shifted by sixteen per cent towards the red end of the spectrum, they all made sense. But what was the origin of the redshift? Normally a redshift in a stellar spectrum implies motion of the star away from us (p. 77), but this redshift was far too great to correspond to the motion of a star in our Galaxy. One possibility was that 3C273 was a very compact object, in which the gravitational field was so strong that an effect of Einstein's General Theory of Relativity called the *gravitational redshift* came into effect. However Schmidt found that this explanation did not hang together. He concluded that the redshift was simply due to the expansion of the universe discovered by Edwin Hubble (see chapter 21), so that 3C273 was simply receding from us at sixteen per cent of the speed of light. As we shall see, the recession speed of a galaxy in the universe increases in proportion to its distance from us. The recession speed of 3C273 was large compared with those found for most galaxies at that date, therefore 3C273 was incredibly distant, two thousand million light years away. It must also be very luminous, a hundred times more luminous than our Galaxy. Yet the stellar appearance of the object on a photographic plate meant that this huge luminosity was coming from a very small volume. Schmidt's conclusions were published in the journal *Nature* on 16 March 1963, along with the paper by Hazard and his colleagues on their occultation observations and identification of the 3C273 'star'. Other astronomers who had taken spectra of radio stars were immediately able to make sense of their spectra too. Most of them had even larger redshifts than 3C273. In particular, Jessie Greenstein and Thomas Matthews found that the previously incomprehensible spectrum of 3C48 could be understood if it was receding at thirty-seven per cent of the speed of light and their paper announcing the result also appeared in the 16 March issue of *Nature*.

These remarkable objects were given the rather elaborate name 'quasi-stellar radio sources' and this was soon contracted to *quasars* by Hong-Yee Chiu of Columbia University, New York. As some of them were found to vary their light output on a time-scale of weeks or months, they appeared extremely paradoxical. A hundred times as much light as in our whole Galaxy was being generated in a volume not much larger than the solar system. Many astronomers found this too hard to swallow and strenuously searched for other explanations. Popular for a while was the idea that the redshift was indeed due to rapid recession of the objects, but that they were comparatively local objects which had been ejected at speeds near the

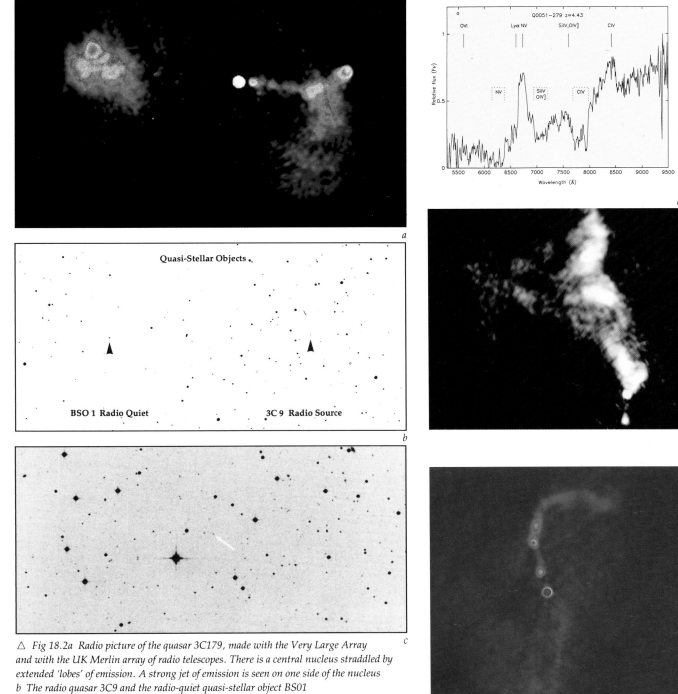

△ Fig 18.2a Radio picture of the quasar 3C179, made with the Very Large Array and with the UK Merlin array of radio telescopes. There is a central nucleus straddled by extended 'lobes' of emission. A strong jet of emission is seen on one side of the nucleus
b The radio quasar 3C9 and the radio-quiet quasi-stellar object BS01
c The most distant quasar known, 0051–279, discovered in 1987
d The spectrum of the quasar 0051–279, with spectral lines shifted in wavelength by 441%. The light from the quasar was emitted when the age of the universe was only 8% of its present age
e Radio map of 3C48 made with a world-wide network of radio-telescopes. The flow becomes turbulent soon after leaving the core of the quasar (spot at bottom right)

△ Fig 18.3 Radio picture of the quasar 2300–189. Two curved jets, each half a million light years long, reach out into space from the central quasar

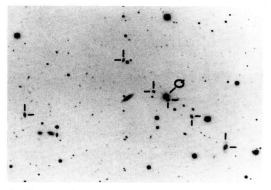

△ *Fig 18.4 The quasar, 2135–147, in a cluster of galaxies at the same distance*

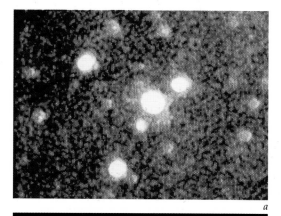

a

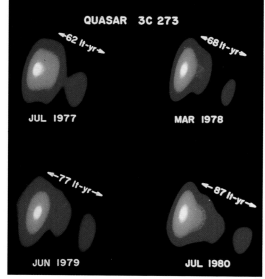

b

△ *Fig 18.5a The quasar 3C275.1 the first to be found at the centre of a cluster of galaxies*
b A sequence of radio images of the quasar 3C273, illustrating the apparent faster-than-light motion of the radio blobs

speed of light from some nearby galaxy. It soon became clear, however, that the energy needed for this ejection was greater than that required if quasars' redshifts were assumed to be due to their being at great distances in an expanding universe.

When I started work as a postgraduate student with William McCrea in October 1964, one of the first problems he gave me to investigate was the distribution in the sky of the quasars. He had noticed that the first dozen or so to be discovered all seemed to lie in one plane. As more quasars were discovered, it soon became apparent that they were distributed fairly smoothly round the sky. A friend of mine, Michael Penston, and I wrote a paper in *Nature* in 1966 in which we refuted claims to the contrary, and noted that the only rather sparsely populated part of the sky corresponded to the summer sky in the northern hemisphere. We attributed this, rather scurrilously, to the absence of observers during the conference season.

Although the distribution of quasars round the sky turned out to be quite smooth, as would be expected for a population of objects spread at random through the universe, their distribution with depth was more interesting. Because quasars were being found with a wide range of redshifts, and hence distances, we were also sampling them at very different epochs in the history of the universe. In 1968 it became clear to me, and simultaneously to Maartin Schmidt, that the quasar population had changed its properties dramatically over the range of cosmological epochs which they spanned. Taking into account the counts of radio-sources to very faint fluxes then available, I concluded that the typical quasar was much more luminous in the past. Maartin Schmidt preferred the interpretation that there were simply many more quasars in the past. While Maartin's view prevailed for more than a decade, it looks as though recent studies have shown my view to be nearer the mark. Although we still do not know why the quasar population was more luminous in the past, it looks as if interactions and mergers between galaxies may have played an important role.

Radio maps of quasars showed that they often had a double structure, like those of radio-galaxies. With time it has come to be seen that quasars are virtually indistinguishable in their radio properties from luminous radio-galaxies. When optical searches for quasars began to be mounted, it became apparent that the radio quasars are only a small proportion of the total population. The majority of quasi-stellar objects are extremely weak radio emitters and so the name quasar, with its connotations of radio-sources, was no longer appropriate. The optically detected objects became known as *quasi-stellar objects*, or simply QSOs. The optical spectra of QSOs (and quasars) showed that the emission came from a hot, dense cloud of gas moving around at speeds up to 10,000 kilometres per second. In this they resemble a class of galaxy discovered by Karl Seyfert in the 1940s, whose nuclei have similar spectra, but the power output of a Seyfert nucleus is on a much smaller scale than many of the quasars. Gradually the properties of quasars seemed less outlandish, compared with other active galaxies like radio-galaxies and Seyferts, than had at first been thought. The controversy about the distances of quasars and QSOs simmered on for many years, though, mainly fuelled by the remarkable claims of Halton Arp. He believed that many or even most quasars are associated with foreground galaxies and he managed to find some highly suggestive

a

b

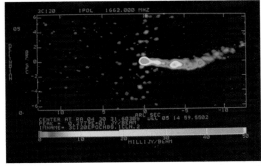

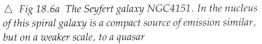

c

△ *Fig 18.6a The Seyfert galaxy NGC4151. In the nucleus of this spiral galaxy is a compact source of emission similar, but on a weaker scale, to a quasar*

b The Seyfert galaxy NGC 1275 photographed in the red light of hydrogen (H-alpha). The extensive system of long filaments appears to be material exploding outwards into space at 1500 miles per second. The galaxy is also a strong radio and X-ray source

c The radio-galaxy 3C120 in which, as in several quasars, the radio spots appear to move outwards faster than the speed of light. This is believed to be an effect of material moving towards us close to the speed of light and near to our line of sight

examples of such associations. For a while I was impressed by Arp's evidence and I explored the idea that quasars, like the nebulae before them, might contain two quite distinct populations, one relatively nearby and the other at cosmological distances. The clinching evidence that quasar redshifts are simply due to the expansion of the universe was the discovery that many quasars are in groups of galaxies with the same redshift as the quasar. Alan Stockton of the University of Hawaii has compiled an impressive dossier of such cases.

The picture that emerged to explain both quasars and double-lobed radio-galaxies was that of a very massive black hole into which matter, gas and perhaps whole stars, is falling. The matter collects into a disc circulating at high speed around the black hole, analogously to the accretion disc in X-ray binaries and dwarf novae (chapter 10). The matter at the inner edge of the disc is moving close to the speed of light and is very hot, emitting at X-ray wavelengths. At the poles of the rotating black hole a beam forms in two opposite directions and particles are accelerated along this beam at speeds close to that of light. These beams impinge on gas surrounding the galaxy and create those huge lobes of radio emission. It is possible that the most violently variable sources are those where our line of sight is such that we are looking straight down one of these beams.

A fascinating development has been the discovery in recent years of several cases where two or more identical quasars are located very near each other on the sky. These seem to be due to another effect of General Relativity, the *gravitational lens*. A galaxy located along the line of sight between us and the quasar gravitationally bends the light from the quasar round it, acting like a lens. The result is that two or more images may be seen around the direction of the lensing galaxy. The 'lens' also magnifies the brightness of the background quasar. In most cases the lensing galaxy cannot be seen, but there are several cases where the lensing galaxy has been identified.

Alone among the objects which form the focus of this book, 3C273 was not known two hundred years ago. Of course we have not understood the true nature of most of the objects until this century, and in many cases, until the past twenty years. But quasars are totally a product of the new astronomies. Perhaps if Jansky had not stumbled on the radio emission from the Milky Way in 1933 and we were still restricted to observations in the visible band, we might by now have come across this stellar object and realized something of its significance. But the beautiful radio structures illustrated in the past two chapters would have remained hidden. The unravelling of the mystery of 3C273 has been a fascinating story, and is probably not over yet.

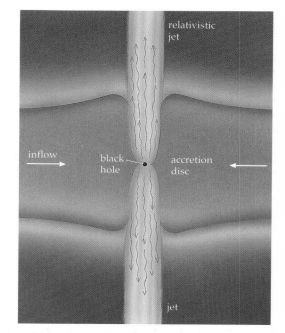

△ *Fig 18.7 Model for central regions of quasar, showing a massive black hole surrounded by an accretion disc, which radiates at X-ray and ultraviolet wavelengths. Twin, oppositely-directed, relativistic beams emerge along the axis of the disc and create the characteristic double radio-source*

▷ Fig 18.8a *Radio picture of the remarkable gravitational lens quasar 0957+561. The bright image at the bottom and the bright image at the top straight above it are both gravitational lens images of the quasar. The faint image just above the lower bright image is the galaxy which is responsible for the lensing. The other patches of emission are part of the extended structure of the quasar not affected by the lens. The colours are coded so that red is bright and green is faint*

b Left-hand frame: optical false-colour picture of the gravitational lens quasar 0957+561. The two gravitational lens images of the quasar are seen, with the lensing galaxy merged into the lower image. Right-hand frame: to bring out the lensing galaxy and to show how identical the two lens images are, the upper image has been subtracted from the lower. The galaxy is now seen clearly in the lower part of the picture. The black ring is an artefact of the photographic processing

c Another remarkable example of gravitational lensing, MG1131+0456, in which the two lensed images are joined by a ring

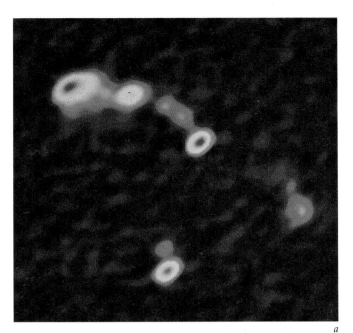

a

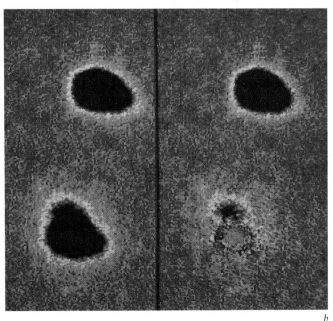

b

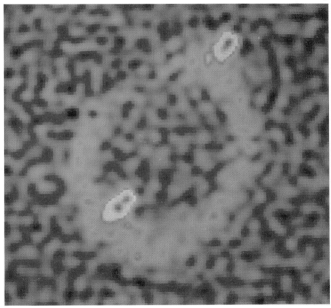

c

C H A P T E R   1 9

A most extraordinary object

WILLIAM PARSONS (LORD ROSSE) *1871*

MESSIER 82

STARBURST GALAXY

Number 82 in the Catalogue of Charles Messier is a peculiar, cigar-shaped galaxy, with a conical bouquet of filaments of gas and dust emerging from its centre. It was first discovered, along with its close companion Messier 81, in December 1774 by J. E. Bode at Berlin. He described it as 'a nebulous patch, very pale, elongated'. Charles Messier, in 1781, thought it 'less distinct than the preceding [i.e. M81]; the light is faint and elongated with a telescopic star at its extremity ...'. A century later, in 1871, William Parsons, Lord Rosse, observed it with his giant reflecting telescope and saw for the first time some of its bright and dark detail: 'A most extraordinary object, at least 10 arcminutes in length and crossed by several dark bands'.

◁ *Fig 19.1 The irregular cigar-shaped galaxy Messier 82*

Messier 82 is a peculiar object, with some resemblance to an edge-on spiral galaxy. In a large telescope, however, it appears to be an elongated, amorphous mass, fading into hazy wisps and filaments around the edges, and heavily spotted with irregular streaks and clumps of obscuring dust. In 1963 Roger Lynds of Kitt Peak National Observatory and Allan Sandage of Mount Wilson Observatory studied M82 in detail, observing it in the light of the H-alpha hydrogen line to emphasize the hot gas in the galaxy and its motions. They found a vast cone of gaseous filaments emanating from the centre of the galaxy, which appeared to be moving at almost a thousand kilometres per second away from the galaxy. They concluded that the galaxy had undergone an enormous explosion and linked this to the fact that M82 is a strong radio source. The cone of filaments was supposed to be the debris from the explosion. Shortly after this, Geoffrey and Margaret Burbidge and Allan Sandage wrote a review article in which they concluded that M82, the nuclei of Seyfert galaxies, M87, Cygnus A, Centaurus A and other radio-galaxies, were all examples of galaxies in which violent explosions had occured in their nuclei. The earlier idea of Baade and Minkowski that collisions between galaxies were the cause of radio-galaxies was swept away, and the concept of violent events in galactic nuclei dominated for over a decade.

Over the past twenty years the picture has gradually changed yet again.

▷ *Fig 19.3 Radio picture of M82. The spots of emission are the remnants of supernova explosions, showing that there has been an unusual number of massive stars forming in this 'starburst' galaxy*

▷ *Fig 19.4 A radio picture of the two companions of M82, Messier 81 (upper right) and NGC 3077 (left), showing clear evidence of interaction between them. The gas seen at the bottom of the picture has been pulled out of one of the galaxies by tidal forces*

▽ *Fig 19.2 Map of the hydrogen gas in Messier 82, made in the characteristic wavelength of the 'H-alpha' emission line. Note the conical shaped 'bouquet' of gas filaments spreading out from the centre of the galaxy*

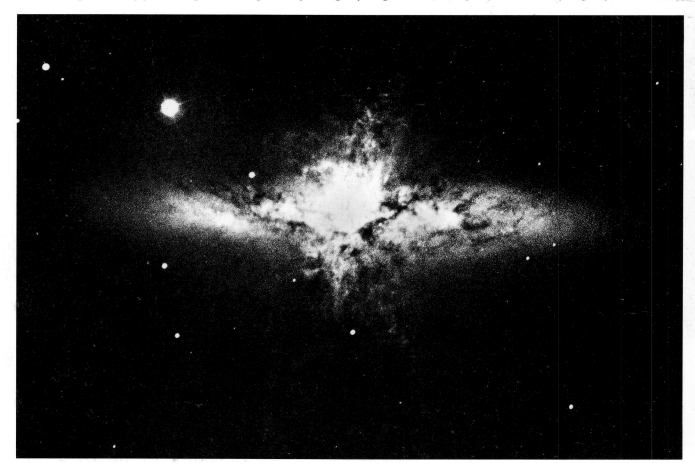

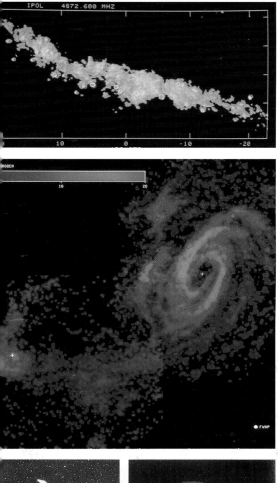

In 1969 Douglas Kleinman and Frank Low of the University of Arizona found that M82 was an exceptionally strong infrared source. At first this was thought to be caused by the same galaxy explosion in the galaxy's nucleus, but with time the explanation has become more mundane, namely that there had been a sudden burst of star formation in the central regions of the galaxy which was shrouded from view by dust. Could the filaments represent material flowing *inwards* towards the centre of the galaxy and fuelling the star formation? This was an idea explored by Philip Morrison and his colleagues from Massachusetts Institute of Technology in 1977. Their idea was that the galaxy M82 was running into a cloud of gas, which was being channelled to the nucleus to form stars.

Messier 82 is a member of a small group of galaxies rather like the Local Group of which our Galaxy is a member (p.131). The two most prominent galaxies in the group are Messier 81 and 82 and the group is called the M81 group, one of the closest to the Local Group. Perhaps the gas flowing into the nucleus of M82 is the result of gravitational interactions between the members of the M81 group. Radio observations, which reveal a huge cloud of hydrogen gas between the galaxies of the group, seem to confirm this idea. Recent radio maps have revealed dozens of supernova remnants in the centre of M82 and this again lends strong support to the idea of an enormous burst of star formation.

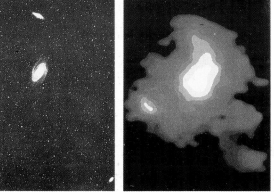

△ *Fig 19.5 Radio map of hydrogen in the M81–M82 system, compared to an optical picture of the same system. M81 is at the centre, M82 above and NGC3077 to the left. The huge cloud of hydrogen around the three galaxies suggests strong tidal interaction between them*

▷ *Fig 19.6 Close-up of M81 at optical wavelengths*

M82 is at least ten times more powerful in the infrared than our Galaxy, which shows the exceptional nature of the star formation activity going on there. This has led to the name *starburst* galaxy for such galaxies. Another example, also a member of a group of galaxies close to the Local Group, is NGC 253. The letters NGC stand for New General Catalogue, a catalogue of nebulae prepared at the end of the nineteenth century, based primarily on the surveys of Caroline and William Herschel and his son John.

Even before the launch of the IRAS satellite, over a hundred starburst galaxies had been identified at visible and ultraviolet wavelengths by their extremely blue colour and characteristic optical spectra with bright emission lines, both signs of massive star formation. The galaxy NGC1068 was an especially interesting example. Detailed mapping of this galaxy at visible wavelengths, and at the infrared wavelengths available to ground-based telescopes, showed both a compact, quasar-like, Seyfert nucleus and a burst of star formation taking place in a ring about 10,000 light years in radius. The starburst is now thought to be linked to the bar seen across the disc of the galaxy, which channels gas into its central regions.

The advent of the IRAS all-sky infrared survey marked a dramatic advance in the importance of the starburst phenomenon. The survey has found tens of thousands of starburst galaxies, some at distances greater than a thousand million light years. Especially remarkable are the many dozens of examples now known whose total infrared luminosity is more than a hundred times that of our Galaxy, in other words as luminous as the most luminous quasars. In a survey of 2400 IRAS galaxies spread over the whole sky, my collaborators and I have found over a hundred of these monsters. Can these still be due simply to extraordinarily powerful bursts of star formation, or must some more exotic phenomenon like giant black holes be invoked?

One of the first examples of these superluminous starburst galaxies to turn up was Arp 220, one of the brightest infrared galaxies in the sky apart from the Local Group galaxies, M82 and NGC253, yet far more distant and hence far more luminous. It appears in Halton Arp's *Atlas of Peculiar and Interacting Galaxies* and is now believed to be an example of a merger between two galaxies. Many others of these superluminous galaxies, almost all according to some astronomers, are also the result of interactions or mergers between two galaxies.

▷ *Fig 19.7 The starburst galaxy NGC253*

▽ *Fig 19.8a Optical photo of the unusual galaxy NGC1068, which has a Seyfert nucleus (the central few hundred light years) surrounded by a ring of active star formation 10,000 light years in radius b The ring of star formation in the centre of NGC1068, seen in the light of the H-alpha line of hydrogen*

a

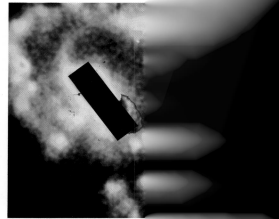

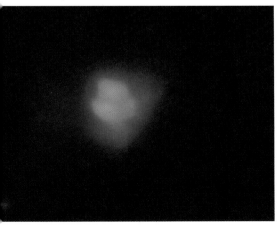

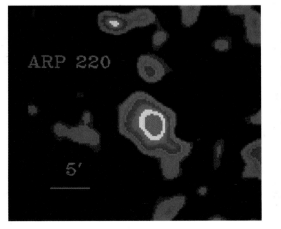

◁ Fig 19.9 *The luminous infrared galaxy Arp 220, in visible light (left) and X-rays (right). It is thought to be the result of a merger between two galaxies, but the heavy obscuration by dust makes it hard to see what is going on*

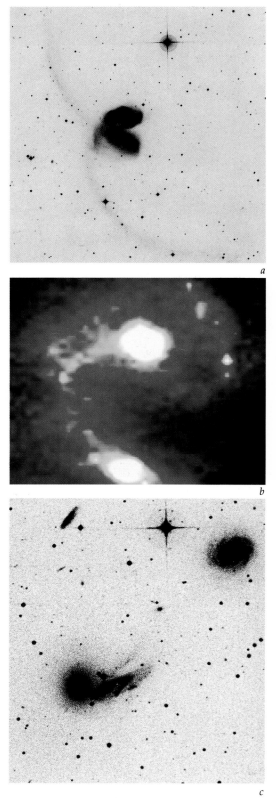

a

b

c

When two large galaxies approach each other, their mutual gravitational attraction starts to modify the shapes of the galaxies in a spectacular way. Plumes and filaments are torn off the galaxies, which take up a bewildering variety of shapes. Some of these have been given nicknames by astronomers (the 'Antennae' galaxies, the 'Fly's Wings' and so on). Computer simulations of the interaction between two galaxies are able to reproduce some of these strange features. Over a period of time the larger galaxy may completely swallow the smaller; as we saw earlier, our Galaxy is in the process of breaking up and swallowing the Magellanic Clouds (p. 121).

Astronomers are creatures of fashion. One year everything unexplained is an explosion in a galactic nucleus, a few years later it will be due to material falling towards a massive black hole, and then suddenly every unusual phenomenon is an interaction between galaxies. It is clearly embarrassing that Baade and Minkowski's idea of collisions between galaxies was neglected for so long. The idea that interactions and mergers are the explanation of luminous starbursts, and perhaps even provide the fuel for quasars, is an interesting one, but may not be the whole story.

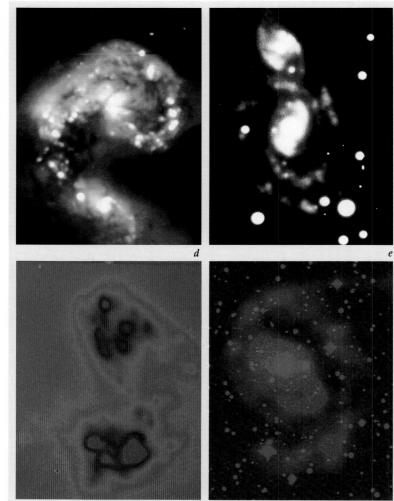

d

e

f

g

◁ *19.10 Examples of interacting galaxies:*

a The 'Antennae' galaxies. These galaxies are believed to be in the process of a head-on collision which began 500 million years ago

b Near infrared image of the 'Antennae' galaxies, showing the distribution of stellar mass in the two galaxies

c 'Fly's Wing' (lower left): an elliptical galaxy has passed through the centre of a spiral a few hundred million years ago, with devastating effects on the spiral

d A colour-enhanced view of the Antennae galaxies which shows intricate dust structure and the locations where new stars are forming

e NGC2207 and IC2103. The smaller galaxy above has a ring of star formation due to the close encounter with the larger spiral galaxy, NGC2207

f False-colour image of the interacting galaxies NGC3690 and IC694

g Markarian 348, a tidally disturbed Seyfert galaxy, at radio wavelengths

h NGC2623, interacting galaxies with long streamers

i The interacting galaxies NGC2992 and NGC2993

j The 'Cartwheel', a galaxy with an unusual ring of star formation

k NGC6027, a group of peculiar galaxies interacting with each other

l The 'Playing Mice' galaxies

m Computer simulation of the interaction between two galaxies, by Alar and Juri Toomre. Long filaments are drawn off the galaxies, like those seen in many actual interacting pairs

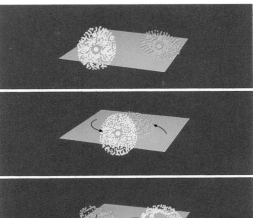

l

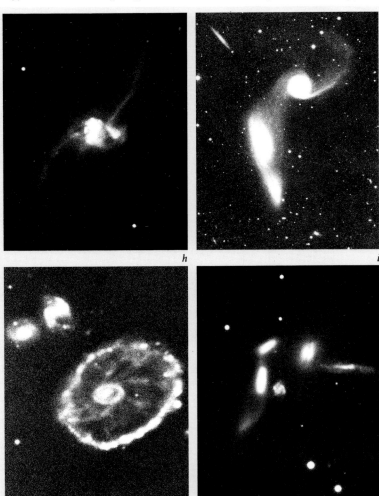

h

i

j

k

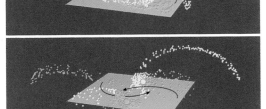

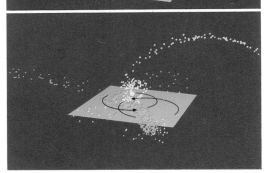

m

CHAPTER 20

... the perceptible Universe exists as a cluster of clusters,
irregularly disposed

EDGAR ALLAN POE *Eureka* (1848)

THE VIRGO CLUSTER

RICH CLUSTER OF GALAXIES

We have come almost to the end of our survey of the contents of the universe. We live in a universe of galaxies, within a large but relatively normal spiral galaxy. The one remaining type of structure we should explore is the giant cluster of galaxies, of which the nearest example lies across the constellations of Virgo and Coma Berenices. The Virgo cluster was first remarked on by William and Caroline Herschel at the end of the eighteenth century. In his 1784 paper on 'the Construction of the Heavens', William Herschel wrote:

> Another stratum (of nebulae) . . . is that of Coma Berenices, as I shall call it . . .
> It has many capital nebulae very near it; and in all probability this stratum
> runs on a very considerable way. It may, perhaps, even make the circuit of the
> heavens . . . the direction of it towards the north lies probably with some
> windings through the Great Bear onwards to Cassiopeia; thence through the
> girdle of Andromeda and the northern Fish, proceeding towards Cetus; while
> towards the south it passes through the Virgin, probably on to the tail of
> Hydra and the head of.Centaurus.

Herschel was describing what we would now call the Local Supercluster (see below). Charles Messier, too, however tiresome he found these nebulae that made comet searching so difficult, must have noticed the concentration of objects towards Virgo in his Catalogue. Like all the objects in this book except the quasar 3C273, the Virgo cluster has been known for at least two hundred years.

The central region of the Virgo cluster contains over three thousand galaxies, concentrated to a volume not much larger than the Local Group of Galaxies, which consists of only twenty galaxies or so. The core of the Virgo cluster is about six million light years across. The distance of the Virgo cluster, like the scale of the universe itself, is still a matter of

△ Fig 20.1 Central part of the Virgo cluster of galaxies, the nearest cluster to us, at a distance of about 60 million light years, including the bright ellipticals M86 (centre) and M84 (right). There are over a thousand galaxies in the dense central core of the cluster and many more in an extended halo (the 'Local Supercluster')

controversy, but I believe we are close to a consensus. Estimates vary from forty to eighty million light years; probably the truth lies in the middle of the range, at about sixty million light years. The two distance estimates which disagree with this most blatantly are the distance derived from supernovae in the Virgo cluster, which is on the high side, and the distance derived from the rotation velocities of spiral galaxies, which is on the low side. This latter distance method is based on a correlation between galaxy luminosity and rotation velocity discovered by Brent Tully and Richard Fisher in 1977. Other methods, using novae in Virgo cluster galaxies, or based on elliptical galaxies, are in good agreement with each other.

Gerard de Vaucouleurs, a French astronomer based at the University of Texas, argued for many years that the Virgo cluster is simply the core of a much larger structure, the Local Supercluster. The Local Group of Galaxies would be part of this Supercluster, lying towards the edge of it. Though de Vaucouleurs' idea was ignored for many years, it has recently come to be accepted. In astronomy the pendulum of fashion swings to and fro. In fact some astronomers believe the Local Supercluster is so vast that it encompasses most of the galaxies within three hundred million light years in a vast irregular disc of galaxy clusters.

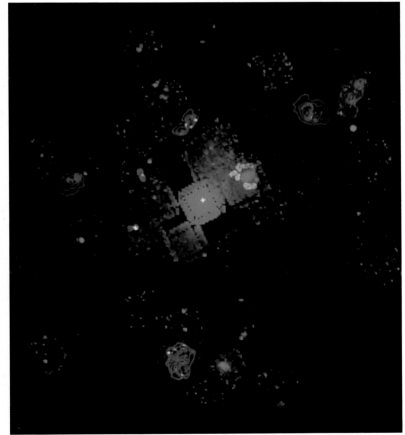

△ *Fig 20.2 A composite view of the Virgo cluster showing X-ray emission (in blue), mainly centred on M87, together with radio maps (in red, expanded for clarity) of the hydrogen in the ten brightest spiral galaxies*

▷ *Fig 20.3 (centre) The giant cluster of galaxies in Coma, about 6 times further away than the Virgo cluster. Whereas Virgo consists mainly of spiral galaxies, Coma has far more elliptical and lenticular galaxies. It is also a powerful source of X-rays, emitted by a huge cloud of gas at a temperature of 100 million degrees centigrade*

a

b

▽ Fig 20.4 (inset) Other prominent clusters of galaxies:
a   The Fornax cluster
b   Cluster number 1060 from the Abell Catalogue
c   The Centaurus cluster
d   The cluster 'CA 0340–538'

c

d

There are several other very prominent rich clusters within five hundred million light years, many of which have been given the names of the constellations where they are found: Coma, Perseus, Pisces, Hydra, Centaurus and Hercules. These are all on the scale of the Local Supercluster or larger. In 1958 George Abell, of the University of California at Los Angeles, made a very valuable catalogue of rich clusters of galaxies in the northern sky, containing over two thousand clusters, some as distant as five thousand million light years. What fraction of galaxies are contained in such clusters depends on how far outwards their haloes extend. If the clusters consist only of the easily observable cores, then only a small percentage of galaxies lie in rich clusters. Some astronomers, myself included, believe that the haloes around these clusters are so extensive that practically all galaxies would be included within them.

The launch of the Uhuru X-ray satellite in 1970 led to a very interesting and surprising discovery about rich clusters. Many of them are strong sources of X-ray emission and this seems to come from vast clouds of very hot gas in between the galaxies, at a temperature of a hundred million degrees centigrade. The gas seems to have been heated up by the action of the galaxies in the cluster streaming through it. It has been found that the gas contains the element iron, identified by its characteristic X-ray emission, and this is taken to imply that at least some of the gas once underwent thermonuclear processing in stars. It was probably once the interstellar gas of spiral galaxies, swept from the galaxies through interactions with their neighbours and the intergalactic gas.

The discovery of gas between the galaxies in clusters poses the question whether there might be gas between the clusters, spread smoothly through the universe. To have avoided detection at radio wavelengths the gas would have to be very hot and so the best hope of detecting intercluster gas is through an X-ray background radiation. Such a background was indeed discovered by the Uhuru satellite and some astronomers believe that it is due to hot intercluster gas, though there is a problem explaining where the energy came from to heat it. Most astronomers believe the X-ray background comes from distant quasars or starburst galaxies, which are known to be X-ray sources.

The existence of clusters of galaxies and of galaxies themselves poses a very interesting question for astronomers and cosmologists. How did they form? Why are the stars not spread uniformly through the universe? Why galaxies are found in clusters has been one of the most hotly investigated topics of theoretical cosmology in the past decade, with a definitive answer still not in sight.

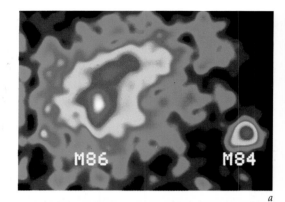

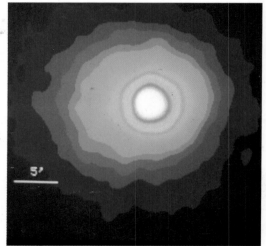

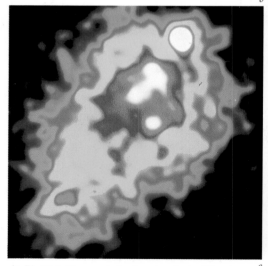

△ *Fig 20.5 X-ray emission from hundred million degree gas in clusters of galaxies:*
a The Virgo cluster
b The Perseus cluster
c The Hydra cluster

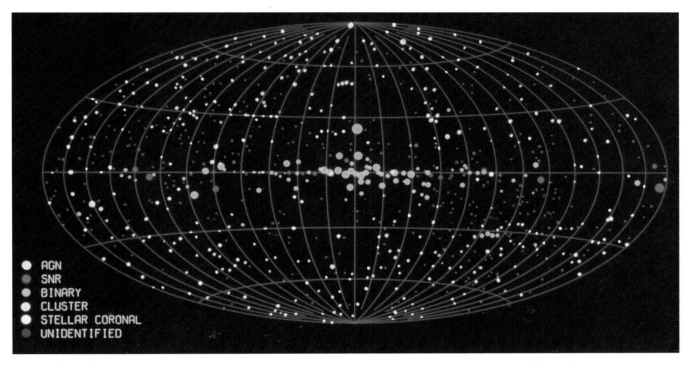

△ Fig 20.6 *The X-ray universe, as surveyed by the HEAO-1 X-ray satellite. Along the Milky Way can be seen the bright X-ray binary sources within our own Galaxy. At higher latitudes the sources are mainly rich clusters of galaxies, with a few 'active' galaxies like Seyferts and quasars*

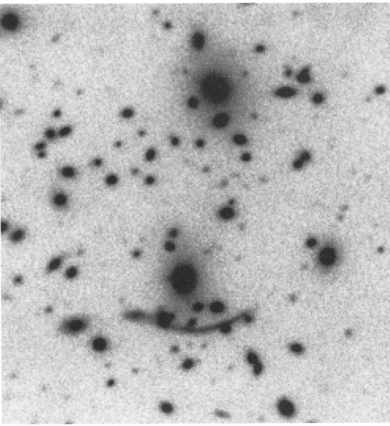

◁ Fig 20.7 *A luminous arc in the distant galaxy cluster Abell 370, studied by Roger Lynds and Vahe Petrosian. It is believed to be a gravitational lens image, made by galaxies in the cluster, of a much more distant galaxy*

C H A P T E R   2 1

Ten billion years before now,
Brilliant, soaring in space and time
There was a ball of flame, solitary, eternal,
Our common father and our executioner.
It exploded, and every change began.
Even now the thin echo of this one reverse catastrophe
Resounds from the furthest reaches.

PRIMO LEVI *'In The Beginning'*

THE UNIVERSE

With superclusters we reach the largest structures we know about at present. On even larger scales the universe starts to look smooth and regular, remarkably so in fact. The earth, the solar system, the Milky Way,

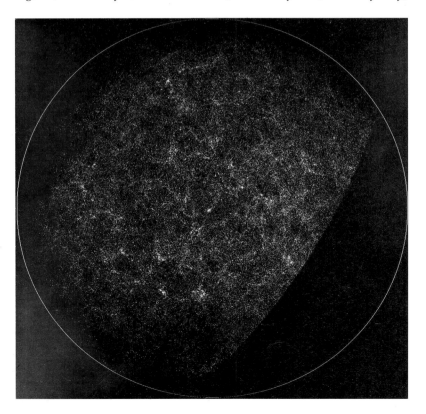

◁ Fig 21.1 A visualization of the distribution of galaxies in the northern sky. Each dot represents many galaxies. The galaxy counts were made by C. D. Shane and C. A. Wirtanen at the Lick Observatory and the visualization was constructed by Jim Peebles and colleagues at Princeton. The tendency of galaxies to congregate in clusters, and the smoothness on the large scale, can be seen

are such striking departures from a smooth distribution of matter, that it is not surprising that geocentric pictures of the universe have dominated human culture for most of recorded history. The suggestion of Democritus in the 5th century BC that the stars are spread smoothly through the universe appears all the more remarkable. Such a picture was discussed by Newton and by the supporters of the island universe theory like Christopher Wren and Immanuel Kant. But only in our century has the truth of this idea become clear.

In 1925 Edwin Hubble made the revolutionary discovery that the universe is expanding:

> The results establish a roughly linear relation between velocities and distances among nebulae for which velocities have previously been published, and the relation appears to dominate the distribution of velocities. . . . The outstanding feature is the possibility that the velocity-distance relation may represent the de Sitter effect, and hence that numerical data may be introduced into discussions of the general curvature of space.

Hubble had found that galaxies are moving away from us with a speed proportional to their distance from us. And though this seems to put us in a special place, at the centre of this expanding universe, it is easy to see that in a smooth enough universe the same would be true for every other galaxy. Whichever galaxy an observer happened to be sitting in, he or she would see the same picture of galaxies receding in every direction.

This concept of an expanding universe took the scientific world almost totally by surprise. Even Einstein, who had had the audacity to propose in 1916 that the universe was on average completely smooth and looked the same at every place and in every direction, did not at first escape from the strait-jacket of a static universe. In 1917 the Dutch cosmologist Willem de

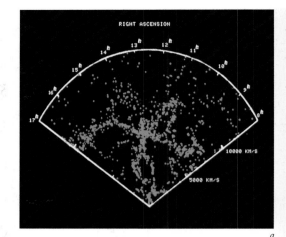
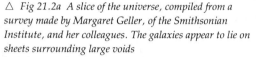

△ *Fig 21.2a A slice of the universe, compiled from a survey made by Margaret Geller, of the Smithsonian Institute, and her colleagues. The galaxies appear to lie on sheets surrounding large voids*
b A deeper galaxy survey made by George Efstathiou, of the University of Oxford, and colleagues, using an automatic plate-measuring machine to measure the position and brightnesses of the galaxies

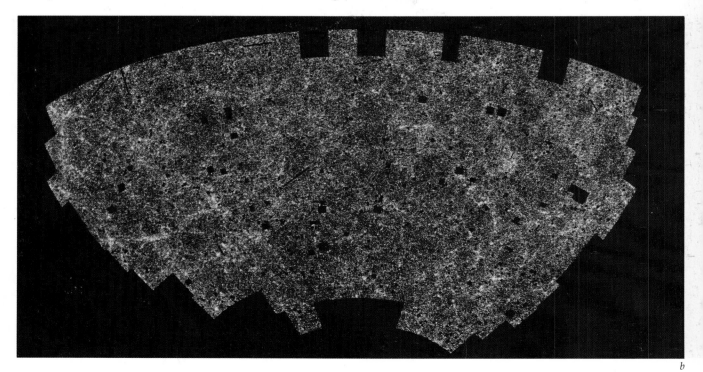

b

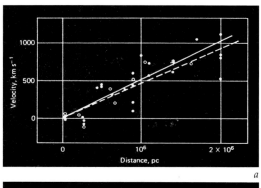

a

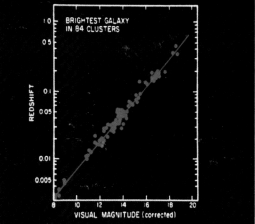

b

△ *Fig 21.3a Hubble's 1929 version of the velocity-distance law. The further away the galaxy, the faster it is moving away from us and the greater its light is redshifted. This was the first convincing evidence of an expanding universe*

b A modern compilation of the Hubble law for bright galaxies in clusters, reaching one hundred times deeper than Hubble's survey

◁ *Fig 21.4 Edwin Hubble, 1889–1953*

The Universe is an infinite sphere, the centre of which is everywhere, the circumference nowhere.

PASCAL *Pensées*

The eternal silence of those infinite spaces strikes me with terror.

PASCAL *Pensées*

Sitter showed that the equations of Einstein's General Theory of Relativity allowed mathematical solutions with an expanding space, in which the light from distant observers would suffer a redshift. This is the 'de Sitter effect' referred to by Hubble. But de Sitter's mathematical 'universe' had no matter in it, and it was by no means clear that it was relevant to a real universe with matter in it. Few astronomers would have been aware of two remarkable papers by the Russian cosmologist Alexander Friedmann, which appeared in 1922 and 1924 in the German journal *Zeitschrift fur Physik* and correctly expounded the possible models for a smooth, ideal universe within the framework of General Relativity. For most people, Hubble's discovery seemed to come out of the blue.

Let us imagine that, like old Qfwfq in Calvino's *Cosmicomics*, we were there at the beginning and made a film of the universe from the earliest time. If we ran this film backwards, what would we see? The galaxies would all start crowding together. But before they started overlapping, they themselves would dissolve away into a smooth sea of gas, still falling together. At some finite time in the past, between ten and fifteen thousand million years ago, the whole universe would be crushed together at infinite density, the moment of the Big Bang.

When Hubble first derived the age of the universe, he obtained much too short an age, only two billion years, shorter even than the age of the earth. Subsequent revisions by Baade in the 1940s and Sandage in 1958 increased this age to a much more reasonable value. The 1970s saw fierce controversy between Allan Sandage and Gustav Tammann on the one hand, and Gerard de Vaucouleurs on the other, about the precise scaling between recession velocity and distance, with a disagreement between them of a factor of two. While the causes of the disagreement are now understood (and explained in my book *The Cosmological Distance Ladder*), there is still controversy, though I believe it is close to resolution.

What was there before the Big Bang? It is curious how this academic question continues to prey on people's minds. A very similar question perplexed the medieval theologians. What did God do before the Creation? Had he passed an infinite time in divine idleness? Saint Augustine of Hippo (AD 354–430) had produced a classic answer, which has never been

Naturally, we were all there, – old Qfwfq said, – where else could we have been? Nobody knew that there could be space. Or time either: what use did we have for time, packed in there like sardines?

I say 'packed like sardines', using a literary image: in reality there wasn't even space to pack us into. Every point of each of us coincided with every point of the others in a single point, which was where we all were. In fact, we didn't even bother one another, except for personality differences, because when space doesn't exist, having somebody unpleasant like Mr Pber Pber underfoot all the time is the most irritating thing.

ITALO CALVINO *Cosmicomics*

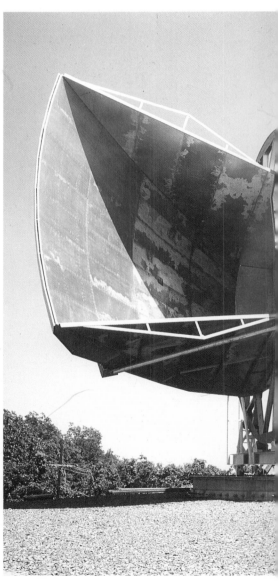

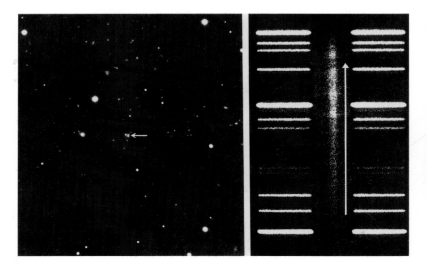

Of course, if he chooses, a person can also take it into his head to find an order in the stars, the galaxies . . .

ITALO CALVINO *Time and the Hunter*

▽ *Fig 21.5 (below left) The Hydra cluster of galaxies and the redshift of the spectrum of one of its galaxies*

▽ *Fig 21.6 (below) The 'Echo' radio antenna with which Arno Penzias and Robert Wilson discovered the microwave background radiation in 1963*

▽ *Fig 21.7 (below right) Albert Einstein, 1879–1955*

bettered: there was no before, time started with the universe. You will find the same answer in Stephen Hawking's *A Brief History of Time*.

What lies outside the universe? There is no outside: either the universe is infinite in extent and always was, or the universe is finite in extent but with no centre and no boundary (this is perfectly possible in a curved space, though hard to visualize). Either way, we will only ever know about a finite part of it, for there is a horizon beyond which we cannot see.

The universe was born in a blaze of light. Radiation dominated all other forms of energy. The realization that we live in a hot Big Bang universe, in which radiation dominated its early history, was one of the great discoveries of this century. The concept was first suggested by George Gamow in 1948 and developed by his students Ralph Alpher and Robert Herman over the next five years. Unfortunately their work attracted little interest at the time. In the early 1960s Bob Dicke and Jim Peebles at Princeton, Fred Hoyle and Roger Tayler in Cambridge, and A. G. Doroshkevich and Igor Novikov in Moscow, were all rediscovering the hot Big Bang model. Quite ignorant of any of this theoretical work, Arno Penzias and Robert Wilson at Bell Telephone Labs were trying to eliminate an unknown source of background emission from a microwave antenna, which had been used in the 'Echo' transcontinental radar experiment. This faint background emission, corresponding to material at a temperature of −270 degrees centigrade, had been known at Bell Labs for some time and presumed to be due to imperfections in the antenna. This temperature is only three degrees above the absolute zero of temperature, −273 degrees centigrade, at which all motions of atoms in matter ceases. Penzias and Wilson painstakingly took the antenna apart, removed what Penzias has referred to as 'a white dielectric substance' left by some nesting pigeons, reassembled the antenna, and flew round it in a helicopter, pointing a transmitter at it to test whether any stray radiation from the ground could reach the radio receiver. But they could not eliminate the emission and concluded that it must be of cosmic origin. Their modest paper, only two pages long, has turned out to be one of the most seminal this century. For this microwave background radiation is a cold whisper left from the days of the Big Bang itself. Today the universe is dark and cold, and getting darker and colder.

The microwave background radiation is the limit of our cosmological knowledge, except for one other thing. Most of the helium in the universe, in the sun, in the earth's atmosphere, in a child's balloon, is a relic of nuclear reactions during the universe's first few minutes of existence. The observed abundances of helium, of deuterium ('heavy' hydrogen) and of lithium, agree with the predictions of theoretical calculations of what should have been produced in the early universe.

The expansion of the universe, the microwave background radiation and the light element abundances all fit well with a simple hot Big Bang universe. But there are some philosophical difficulties with this simple picture which seem to require a totally new picture of the early universe.

The microwave background radiation is remarkable for its uniformity around the sky, except that it looks slightly brighter in one direction and slightly dimmer in the opposite direction. The simplest interpretation of this slight non-uniformity is that our Galaxy (with the Local Group and other nearby galaxies) is moving through space at a speed of about six hundred kilometres per second. Recently, a group of colleagues and myself have used 2400 galaxies found by the IRAS Infrared Astronomical Satellite to map out the distribution of mass in the universe within 300 million light years. We find that the motion of our Galaxy through space can be accounted for simply as the result of the gravitational pull of large galaxy clusters, like those illustrated in the last chapter.

▷ Fig 21.9 A map of the microwave background radiation, colour-coded according to its brightness. The red area is slightly brighter than average because we are moving towards this direction and the blue area is slightly fainter than average because we are moving away from this direction. From this map a speed of the Local Group of Galaxies through space of 600 kilometres per second can be deduced. Apart from this consequence of the motion of the Local Group of Galaxies through space, the microwave background radiation is perfectly smooth and the same in every direction

And there is no object so soft but it makes a hub for the
 wheel'd universe,
And I say to every man or woman, Let your soul stand cool
 and composed before a million universes.

WALT WHITMAN *Song of Myself*

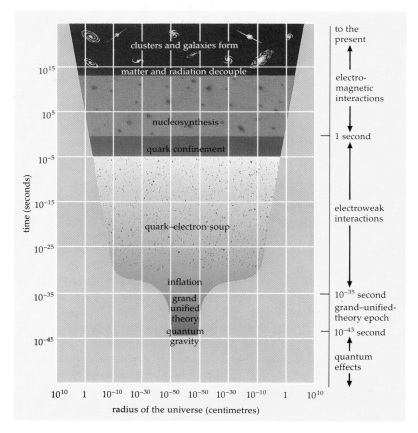

to the present

electro-magnetic interactions

1 second

electroweak interactions

10^{-35} second grand–unified-theory epoch
10^{-43} second

quantum effects

◁ Fig 21.8 Schematic picture of the evolution of the universe. In the 10^{-43} seconds following the Big Bang, quantum effects dominate and the four fundamental forces (electromagnetism, weak and strong nuclear forces and gravity) are believed to have been unified into a single force. First gravity separates out, leaving the other three forces as a 'Grand Unified Force'. When the strong nuclear force separates from the 'electroweak' force 10^{-35} seconds after the Big Bang, inflation begins. The matter in the universe consists of a 'soup' of quarks, which are the building blocks of protons and neutrons, and electrons, but the dominant form of energy is radiation. When the universe is one second old, the quarks bind together to make protons and neutrons, and the weak nuclear force and electromagnetic forces separate. Nucleosynthesis begins and continues until the universe is about three minutes old. When the universe is one million years old, the matter cools sufficiently to become transparent to radiation and galaxies and clusters of galaxies can begin to form.

We have direct evidence for this picture only back to the one second epoch. The extrapolation back in time prior to that epoch becomes increasingly speculative as we approach the instant of the Big Bang

listen: there's a hell
of a good universe next door; let's go –

E. E. CUMMINGS *'Pity this busy master, manunkind'*

Once the motion of our Galaxy through space is allowed for, the microwave background radiation looks the same in every direction to within one part in ten thousand, a remarkable uniformity. Yet the microwave light we are seeing was emitted by matter so far away that the material in the different directions could not have been in any previous communication. How did all these isolated pieces of matter 'know' they had to be the same? This is a real paradox, which in the simple Big Bang model can only be resolved by saying that the universe was born with this uniformity.

This brings us to the second problem with that model. The galaxies, stars and we ourselves demonstrate that the universe is not perfectly uniform. How did these fascinating structures form? In the simple model we again have to resort to assuming that small non-uniformities were present at the birth of the universe, and were of just the right form to evolve into the structures we see today.

Both these problems are solved in a new picture of the early universe, first put forward by Alan Guth of the Massachusetts Institute of Technology, *inflationary cosmology*. In this picture, the whole universe that we see today has inflated from a single infinitesimal 'seed'. The distant matter that we sample with the microwave background radiation was in communication before the inflation period. The energy for the colossal expansion is essentially derived from the gravitational energy of the universe. The trigger for the inflation is the separation of the nuclear and electromagnetic forces into the distinct types of force we see today. Prior to this 'phase transition' there would only have been one 'unified' force (apart from gravity). The detailed physics of the phase transition can also explain the origin of galaxies and why we live in a universe dominated by matter rather than anti-matter.

One requirement of the inflationary model is that the universe contain a very high proportion, perhaps ninety-nine per cent, of dark matter. I mentioned in chapter 14 (p. 113) that the halo of our Galaxy appears to contain a large mass of dark matter, and the same appears to be true for other galaxies. To satisfy the inflationary model, though, there must be an even more all-pervasive distribution of dark matter. Interestingly, the IRAS galaxy survey I described above does point to such a pervasive distribution of dark matter. But what this dark matter could be remains a mystery.

The flowering of many remarkable ideas about the very early universe, Alan Guth's inflationary universe, the 'superstring theory' of my colleague Michael Green at Queen Mary College and others, the attempts to produce a quantum theory of gravity by Roger Penrose, Stephen Hawking and others – all these seem like a wonderful fusion of particle physics and cosmology. Yet they also seem like the disintegration of the science of cosmology, for none of these wonderful ideas makes any direct predictions about our universe, the world in which we live. Let a hundred flowers bloom, be what you want to be, think what you want to think, why should not any idea be as good as any other?

If my book serves any purpose, it is to remind people of the enduring reality of the universe. The stars and galaxies are like hard stones against our feet. The ideas that tell us something new about this universe of ours are the ones that will survive: will survive, that is, if human beings manage not to destroy themselves and their miraculous planet.

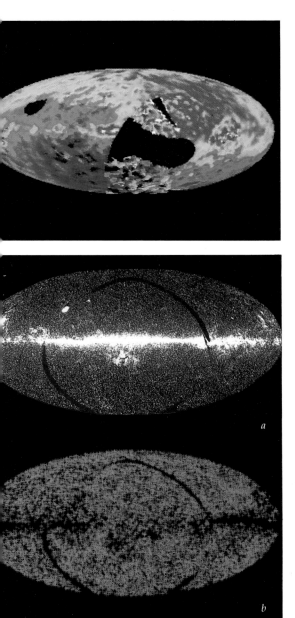

△ *Fig 21.10a The infrared sky. A map of the sources detected by the IRAS infrared astronomy satellite. The yellow sources near the Milky Way are star forming regions in our Galaxy, the blue sources are stars in the Milky Way and the green sources over the whole sky are galaxies*
b The distribution on the sky of the galaxies seen by IRAS. From a detailed study of these galaxies, my colleagues and I have been able to show that the motion of the Local Group of Galaxies through space (see Fig 21.9) is due to the combined attraction of (relatively!) nearby clusters of galaxies like those illustrated in chapter 20

Ah, this universe of light –
such colours, such harmonies
and my mind alive with visions
as I go travelling through all times
past the inscrutable galaxies
floating fire
through the great cities seething with activity
or the dazzling landscapes of summer:
and at my inner ear music
too subtle for air to bear
– this life of the mind, mirror of all earth

ah, this universe of light
so short a life I have
to see your marvels
so few days to walk in the world
so few lungfuls of breath
before the heart stops still
and I float out into that translunar dark

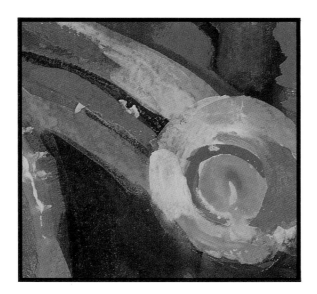

FINDING THE
OBJECTS

Finding the objects described in this book depends on your latitude, the season and time of night. With the aid of a friend who knows the sky or one of the many pocket guides to astronomy with star-maps, available in any good bookshop, you should easily find Sirius, Polaris and the Plough, Vega, Orion, Algol, the Hyades and Pleiades, Delta Cephei, the Andromeda Nebula and the Milky Way, if you live in the northern hemisphere. If you live in the southern hemisphere, you should easily find Alpha Centauri, Sirius, Orion, the Hyades and Pleiades, the Milky Way and the Magellanic Clouds. For the fainter objects you will need to invest in a small telescope or good binoculars and *Norton's 2000.0 Star Atlas and Reference Handbook*. Here are a few simple tips for finding the brighter objects:

Northern mid-latitudes

Facing north at any time of year you will easily find the Plough or Big Dipper, whose end two stars point to Polaris. On the opposite side of Polaris from the Plough/Big Dipper you will recognize the W of Cassiopeia, most favourably observed on an autumn evening. Following the right-hand arm of the W leads you to the kite of Cepheus and Delta Cephei (see Fig 9.1). The right-hand V of the W points towards the Andromeda Nebula (see Fig 16.1 for its relationship to the brighter stars of Andromeda), while the left-hand V of the W points towards Perseus and Algol.

Facing south on a January evening, the constellation of Orion is a spectacular sight. The Nebula is the central star of the sword. Following the line of the three stars of the belt to the left leads to Sirius, the brightest star in the sky. Following the line of the belt to the right leads to Aldebaran, the brightest star in Taurus and the left-hand end of the V of the Hyades cluster (see Fig 3.1). Continuing on the same line for a similar distance leads to the unmistakable Pleiades cluster (Fig 13.1).

Overhead on an August evening you will find the bright star Vega. This is also the best time to see the Milky Way.

Southern mid-latitudes

Facing south you should easily find the Large and Small Magellanic Clouds, at their best in the summer months. This is also the best time to find Orion, Sirius, the Hyades and Pleiades, towards the north. An April evening is the best time to find the Cross, located in the dark Coalsack region of the Milky Way. Following the Cross round the sky are the two bright stars Beta and Alpha Centauri and preceding it at about the same distance is Eta Carinae, at present visible only with binoculars.

FURTHER READING

Star Names, Their Lore and Meaning Richard Hinckley Allen (Dover, 1963). Scholarly compilation of myths and literary quotes about stars and constellations.

Cambridge Atlas of Astronomy ed. J. Audouze and G. Israel (Cambridge University Press, 1985). Heavily illustrated guide to astronomy at a significantly more advanced level than this book.

A Short History of Astronomy Arthur Berry (Dover, 1961). A delightful history of astronomy up to 1900.

Burnham's Celestial Handbook Robert Burnham Jr (3 volumes, Dover, 1978). Comprehensive and inspiring compilation of information about astronomical objects, by constellation.

In Search of the Big Bang John Gribbin (Heinemann, 1986). Highly readable account of modern ideas about cosmology and particle physics.

Astronomers' Stars Patrick Moore (Routledge & Kegan Paul, 1987). Detailed account of what is known about sixteen stars, several of which feature in this book.

Norton's 2000.0 ed. Ian Ridpath (Longman, 1989). Star maps and comprehensive information for the amateur astronomer.

Cosmic Landscape Michael Rowan-Robinson (Oxford University Press, (1979). An introduction to the astronomy of the invisible wavelengths.

The Big Bang Joseph Silk (W. H. Freeman, 1980). Excellent introduction to astrophysics and cosmology.

Pathways to the Universe Francis Graham Smith and Bernard Lovell (Cambridge University Press, 1988). Well-illustrated guide to modern astronomy.

The Invisible Universe Revealed Gerrit Verschuur (Springer-Verlag, 1987). Outline of the radio universe, with black-and-white illustrations.

The First Three Minutes Steven Weinberg (Basic Books, 1977). Very good account of ideas about the early universe (but not including inflation).

New Scientist and *Scientific American* (and many other popular science magazines) regularly contain features on astronomy which will keep you in touch with developments in astronomy.

   If you get really hooked you should join one of the many amateur astronomy societies throughout the world and subscribe to at least one of the magazines aimed at astronomy enthusiasts, such as *Sky and Telescope*, *Mercury* and *Astronomy*.

GLOSSARY

absorption line   dark line on spectrum of source due to absorption of light of particular wavelength by atoms of a particular element

accretion disc   disc of gas accumulated around a compact object (white dwarf, neutron star or black hole)

asteroid   lumps of rock in solar system up to hundreds of miles in diameter, mostly confined to asteroid belt between Mars and Jupiter, but some of which (the Apollo asteroids) travel in orbits which cross that of the earth

Big Bang   explosion of the universe from a state of virtually infinite density, which seems to have occurred ten to fifteen thousand million years ago

binary system   two stars in orbit around each other

black hole   a region in which the matter has collapsed together to such an extent that light can no longer escape from it

Cepheid variable   a kind of star, of which δ Cephei is the proto-type, which pulsates and varies its light output regularly over a period of hours to months. The more luminous the star the longer the period of variation

cluster of galaxies   concentration of hundreds or thousands of galaxies

cosmic rays   charged atomic particles (electrons and atomic nuclei) arriving at the earth at speeds very close to that of light, generated in the sun, in pulsars or during supernova explosions, or in black holes

cosmology   study of the universe as a whole

constellation   group of stars near together on the sky, but not necessarily close together in space, which make up a recognizable pattern

dark matter   matter whose presence is deduced from dynamical evidence but which is not yet detectable through its radiation

Doppler effect   a source of light (or sound) has its wavelength or frequency shifted if it is moving towards us or away from us. Visible wavelengths are shifted towards the red (long wavelength) end of the spectrum if the source is receding, and towards the blue if it is approaching

dwarf star   star with radius smaller than that of the sun. The temperature of the star determines its colour (red for cool stars, white for very hot stars)

dust   most of the atoms heavier than helium in between the stars are in the form of very small grains of dust, from a thousandth to a tenth of a micrometre in radius. The most common types of grain are probably graphite or sooty carbon and silicates

ecliptic   the great circle around the sky on which the sun appears to move through the year as the earth orbits around it. The ecliptic plane is the plane of the earth's orbit. The orbits of the other planets also lie close to this plane

electron   light atomic particle with negative electric charge

emission line   bright line on spectrum of source due to emission of light of a particular wavelength by atoms of a particular element

flare   eruption on surface of star or sun, of magnetic origin

galaxy   giant star system like the Milky Way, of mass ranging from a hundred million to a million million times the mass of the sun. The three main types of galaxy are elliptical, spiral or irregular in appear-ance. Most galaxies have a central concentration of stars and gas, called the nucleus. Spiral galaxies have a disc of gas, dust and young stars. Most large galaxies are surrounded by a halo of old stars and globular clusters

General Theory of Relativity   Einstein's theory of gravitation, in which matter causes space to be curved

giant star   star of radius much larger than the sun. The surface temperature of the star determines its colour, eg. red for cool stars (3000°C), blue for hot stars (10,000°C)

globular clusters   dense aggregations of a million or so old stars, found in the halos of galaxies

gravitational lens   background object, usually quasar, appears as multiple object due to bending of light by intervening galaxy

Hubble law   proportionality of velocity of recession of galaxies and distance, due to expansion of universe

hydrogen-burning   the fusion of hydrogen to make helium, the nuclear reaction which is the power source for the sun and most stars

inflationary universe   theory of very early universe in which whole of present observable universe inflated from minute speck of matter

ionized gas   gas in which most of the atoms have had some of their electrons stripped off either by collision between atoms of gas, if the gas is hot enough, or by energetic ultraviolet or X-ray photons

Local Group of galaxies   group of twenty or so nearby galaxies dominated by our Galaxy and the Andromeda galaxy, and including the Magellanic Clouds

luminosity   total power output of a source of radiation, usually measured in units of the sun's luminosity, which is 4×10^{26} Watts

Messier 1-103   list of fuzzy or nebulous objects compiled by the eighteenth century French comet watcher, Charles Messier

microwave background   highly smooth and isotropic bath of microwave radiation in which we are immersed, believed to be the relic of the fireball phase of the Big Bang

molecular cloud gas cloud in which the most abundant element, hydrogen, is in the form of molecules

nebula fuzzy patch of light on sky which may be gas cloud within our Galaxy or an external galaxy

nuclear reactions processes involving collisions between atomic nuclei which take place at the centres of stars and during the fireball phase of the Big Bang, in which atoms are fused together to make new elements

neutron heavy atomic particle with no electric charge

neutrino neutral, massless atomic particle produced during nuclear reactions at the centre of stars and, profusely, during Type II supernova explosions. They have been detected from the sun and from Supernova 1987A

neutron star tiny kind of dead star only tens of kilometres in diameter, in which the matter is in the form of neutrons crushed together until they touch. They are formed in Type II supernova explosions and are responsible for pulsars

nova star which brightens suddenly by up to a million times, caused by material falling onto a white dwarf from a companion star

open clusters loose aggregates of hundreds of young, massive stars found in the discs of spiral galaxies

parallax apparent motion of a nearby star on the sky as the earth orbits the sun

period-luminosity law relationship between period of variation and luminosity for Cepheid variable stars. Can be used to determine distance of a star system

planetary nebula shell of gas illuminated by hot, compact central star, formed by ejection of the outer layers of a red giant star

photon particle of light. Light can be thought of either as waves or as a stream of particles. A photon carries a definite amount of energy, proportional to the frequency of the light, so the high frequency ultraviolet and X-ray photons carry more energy and have greater penetrating power than lower frequency radio and infrared photons

proper motion relative motion of stars on the sky due to their orbital motion through the Milky Way

proton heavy atomic particle with positive electric charge

pulsar pulsating radio source due to a rotating, magnetized neutron star

QSO/quasar compact source of optical and ultraviolet radiation in the nucleus of a galaxy, of such luminosity that the starlight from the galaxy is hard to see and the source has a stellar appearance on optical photographs. If a QSO (quasi-stellar object) is a strong radio source it is called a quasar, short for quasi-stellar radio-source

redshift the light from all but the nearest galaxies is shifted in wavelength towards longer wavelengths (in the visible band, this means towards the red end of the spectrum). This is believed to be due to the Doppler shift and demonstrates that the galaxies are receding from us – the expanding universe

relativistic particles atomic particles, eg. electrons or protons, moving at speeds close to that of light

Seyfert galaxy a class of galaxy identified by Karl Seyfert in the 1940s in which rapid motions can be seen in the nucleus and with excess ultraviolet radiation compared to a normal galaxy due to the presence of a compact source like a minature QSO

spectrum the spread of wavelengths from a source of light, for example by a prism. Raindrops act like a prism to generate a rainbow, the sun's visible spectrum

starburst strong and sudden episode of massive star formation

supercluster very large cluster of galaxies or cluster of clusters of galaxies

supergiant very luminous giant star

supernova sudden explosion of a star due to either complete disruption of a white dwarf in a binary system (Type I) or to the death of a massive star accompanied by the ejection of the outer parts of the star and implosion of the core to form a neutron star or black hole (Type II)

synchrotron radiation radiation from relativistic particles, especially electrons, spiralling in a magnetic field

variable star star whose light output varies

white dwarf small type of dying star thousands of kilometres in diameter, with a surface temperature of about 100,000°C, in the process of cooling off to a dark, cold, dead state

zodiac the twelve constellations, roughly one for each month, crossed by the ecliptic and in which the sun and planets are always to be found

zodiacal light band of light along the zodiac due to scattering of sunlight by small particles of dust spread through the solar system

ACKNOWLEDGEMENTS
FOR PICTURES

Fig **1.1, 1.4, 8.3, 13.3, 14.7, 20.4a:** David Malin, Anglo-Australian Observatory and Royal Observatory Edinburgh; **1.2:** Dennis Mammano, Astronomical Society of the Pacific; **1.3, 1.5l, 1.6e, 1.11, 2.2, 2.4, 2.9, 6.4, 6.5, 9.6, 12.20, 13.6, 14.2, 14.10a, 15.1, 15.2, 15.3, 15.4, 15.5, 15.8, 16.3, 16.5d, 16.5e, 18.1a, 18.2c, 18.2d, 18.6b, 19.10e, 19.10l, 20.1, 20.3, 20.4c, 20.7:** National Optical Astronomy Observatory, Tucson; **1.5a:** British Museum; **1.5b, 1.15, 11.1, 11.4:** Yerkes Observatory; **1.5d:** Trinity College, Cambridge; **1.5e:** Scrovegni Chapel, Padua; **1.9a, 1.9b, 1.9d, 10.8:** European Space Agency; **1.5f, 1.5i, 1.10, 1.13:** David Hughes and Carole Stott; **1.5g, 4.9, 7.4, 8.2, 10.1, 13.2, 14.5:** Royal Astronomical Society; **1.5h:** National Portrait Gallery; **1.5j:** Lady Lindsay-Finn and the Air and Space Museum The Smithsonian, U.S.A.; **1.5k:** Astronomische Nachrichten; **1.6a:** Patrick Moore; **1.6b:** Tate Gallery; **1.6c:** Graham Leaver; **2.1:** John Laidlaw; **3.1:** Robin Scagell; **4.8:** Colin Taylor; **5.1:** Ken Phillips; **6.1, 8.1, 9.1, 13.1:** Raymond Livori; **16.1:** David Early; **1.6d, 7.2, 9.5, 11.3, 12.19, 18.2b, 18.6a, 19.2, 19.10h, 19.10k, 21.5:** Mount Palomar Observatory; **1.9c:** H. U. Keller, Max Planck Institut fur Aeronomie, Landau, FRG; **1.12, 1.16, 5.3, 5.4, 6.8b, 8.9, 8.13, 8.15, 8.16, 13.5, 14.9f, 14.10b, 14.10c, 15.6, 15.7, 16.4a, 16.4b, 21.10a:** IPAC, California Institute of Technology; **1.14:** Glen MacPherson & Roy Lewis, University of Chicago; **2.6:** L. Golub and K. Topka, Center for Astrophysics, Cambridge, Mass.; **3.2, 3.8, 4.1:** Robert Burnham Jr, Dover Publications; **3.3, 4.4, 7.1, 7.3a, 7.3b, 8.5, 8.10, 10.2, 11.7, 12.1, 12.22, 13.9b, 13.9c, 13.9d, 16.2a, 16.6a, 16.6b, 17.1, 17.4a, 19.1, 19.6, 19.8a:** Lick Observatory; **4.2, 6.6, 7.3c, 7.3d, 8.4, 8.11, 8.12, 8.17a, 8.17b, 8.17c, 12.25, 13.7, 13.9a, 14.1, 15.10, 16.5a, 16.5b, 16.5c, 16.6c, 16.6d, 17.2b, 17.2c, 17.5a, 19.7, 19.10i, 19.10j, 20.4b, 20.4d:** Anglo-Australian Observatory; **4.6:** Ronald Sheridan, Ancient Art and Architecture Collection; **4.10:** British Library; **4.12:** Musee Nationaux, Paris; **5.2:** NASA Ames Research Center; **5.5:** Perkin-Elmer Space Science Division; **12.9, 12.12, 12.17:** Fred Seward, Center for Astrophysics, Cambridge, Mass.; **7.5, 9.2, 14.3, 14.4, 14.8, 21.4, 21.7:** Astronomical Society of the Pacific; **7.6a, 7.6b, 8.7, 8.18a, 8.19, 10.4, 10.9, 12.5, 12.8, 12.21, 14.10e, 14.10f, 16.7a, 16.7b, 17.3b, 17.4b, 17.4c, 17.5b, 17.6a, 17.6b, 17.6c, 17.6d, 17.6e, 17.6f, 17.6g, 17.8, 18.3, 18.5a, 18.8a, 19.3, 19.10f, 19.10g, 20.2:** National Radio Astronomy Observatory, Green Bank; **7.7, 11.9, 17.3c, 18.1b, 18.2a, 18.5e, 18.6c, 19.5:** Peter Wilkinson, Jodrell Bank; **7.8, 16.7c, 19.10b:** Ian McLean; **8.6:** R. Maddelena, Columbia University; **8.8:** Mark McCaughrean; **8.14, 12.27, 19.10a, 19.10c:** Royal Observatory Edinburgh; **8.18b:** R. D. Gehrz, University of Wyoming; **8.21:** Charles Lada, University of Arizona; **6.8a, 9.4, 10.7, 10.10, 12.2, 12.7, 14.9d, 16.4b, 17.3d, 17.4d, 17.5c, 18.1c, 19.9b, 20.5a, 20.5b, 20.5c, 21.2a, 21.3b:** C. Jones, C. Stern and W. Forman, Center for Astrophysics, Cambridge, Mass.; **10.5a, 10.5b:** Center for Astrophysics, Cambridge, Mass.; **10.6:** Lois Cohen, Griffith Observatory; **10.11:** W. K. Hartmann, University of Arizona; **11.2:** Joseph Needham Institute, Cambridge; **11.8:** Gerard de Vaucouleurs, University of Texas; **12.3:** W. C. Miller; **13.4,** **16.2b:** Mount Wilson Observatory; **12.4:** Paul Murdin, Royal Greenwich Observatory; **12.6, 12.16:** David Malin, Anglo-Australian Observatory; **12.10:** Sidney van den Bergh, University of Toronto; **12.13, 12.14:** Jocelyn Bell-Burnell, Royal Observatory Edinburgh; **12.23:** European Southern Observatory; **12.24:** Ian Shelton; **12.26:** F. Reines and J. C. van der Velde, IMB; **13.8:** J. B. Marling; **14.9a:** T. Dame: Center for Astrophysics, Cambridge, Mass.; **14.9b:** Lund Observatory; **14.9c:** C. G. T. Haslam, Max Planck Institut fur Radioastronomie, Munich; **14.9d:** D. MacMannon; **14.10d:** David Allen, Anglo-Australian Observatory; **16.3:** California Institute of Technology; **16.4c:** Elias Brinks; **16.4d, 16,4e:** Robert Braun; **16.7d:** E. R. Deul and J. M. van der Hulst, Westerborg Synthesis Radio Telescope; **16.7e:** J. Kamphuis, J. M. van der Hulst and R. Sancesis, Westerborg Synthesis Radio Telescope; **17.3a:** J. Biretta, Center for Astrophysics, Cambridge, Mass.; **17.5d:** J. O. Burns, University of New Mexico; **18.5b:** T. Pearson, Owens Valley Observatory; **18.4, 18.8b:** Alan Stockton, University of Hawaii; **18.8c:** Bernie Burke, MIT; **19.4:** J. M. van der Hulst and R. I. Allen, Westerborg Synthesis Radio Telescope; **19.8b:** G. Cecil, Princeton; **19.9a:** R. E. Schild; **19.10d, 19.10f:** William Keel, University of Arizona; **20.6:** Kent Wood, Naval Research Labs; **21.1:** P. J. E. Peebles, Princeton University; **21.2b:** George Efstathiou, Oxford University; **21.3a:** Edwin Hubble, 'The Realm of the Nebulae', Dover Publications 1958; **21.6:** Arno Penzias, AT&T Bell Labs; **21.9:** David Wilkinson, Princeton University; Paintings by Hanife Hassan O'Keeffe; Figure artwork by Oxford Illustrators.

SOURCES FOR QUOTES

p. 18 Peter Russell 'All for the Wolves' Anvil Press 1984; **p. 20** R. M. Rilke 'Sonnets to Orpheus', translated by J. B. Leishman, Hogarth Press 1957; Robert Frost 'Complete Poems' 1970; **p. 21** Elizabeth Jennings 'Collected Poems' Carcanet Press; **p. 22** T. S. Eliot 'Collected Poems' Faber & Faber 1963; **p. 26, 38** W. H. Auden 'Collected Poems' Faber & Faber 1969; **p. 32, 74, 110** Osip Mandelstan 'Selected Poems', translated by Clarence Brown and W. S. Merwin, Penguin Books 1977; **p. 41, 122** Italo Calvino 'Mr Palomar', translated by William Weaver, Picador 1986; **p. 44** Sylvia Plath 'Colossus' Faber & Faber 1960; **p. 47, 105** Sappho quotes from 'Greek Lyric Poetry', translated by Willis Barnstone, Bantam Books 1962; **p. 71** R. M. Rilke 'Duino Elegies, translated by C. F. MacIntyre, University of California Press 1961; **p. 80, 164** Primo Levi 'Shema', translated by Ruth Feldman and Brian Swann, Menard Press 1976; **p. 100** George Seferis 'Collected Poems', translated by Edmund Keeley and Philip Sherrard, Jonathan Cape 1969; **p. 103** Benjamin Britten/Montagu Slater 'Peter Grimes', Boosey and Hawkes 1945; **p. 168** Italo Calvino 'Cosmocomics', Abacus 1982; **p. 169** Italo Calvino 'Time and the Hunter', translated by William Weaver, Abacus 1983.

INDEX